T0230476

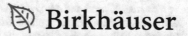

Computer Science Foundations and Applied Logic

Computer Science Foundations and Applied Logic is a growing series that focuses on the foundations of computing and their interaction with applied logic, including how science overall is driven by this. Thus, applications of computer science to mathematical logic, as well as applications of mathematical logic to computer science, will yield many topics of interest. Among other areas, it will cover combinations of logical reasoning and machine learning as applied to AI, mathematics, physics, as well as other areas of science and engineering. The series (previously known as *Progress in Computer Science and Applied Logic*) welcomes proposals for research monographs, textbooks and polished lectures, and professional text/references. The scientific content will be of strong interest to both computer scientists and logicians.

Kenneth J. Supowit

Algorithms for Constructing Computably Enumerable Sets

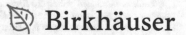 Birkhäuser

Kenneth J. Supowit
Department of Computer Science
and Engineering
The Ohio State University
Columbus, OH, USA

ISSN 2731-5754 ISSN 2731-5762 (electronic)
Computer Science Foundations and Applied Logic
ISBN 978-3-031-26906-6 ISBN 978-3-031-26904-2 (eBook)
https://doi.org/10.1007/978-3-031-26904-2

This book is published under the imprint Birkhäuser, www.birkhauser-science.com by the registered
company Springer Nature Switzerland AG
The registered company address is: Gewerbestrasse 11, 6330 Cham, Switzerland

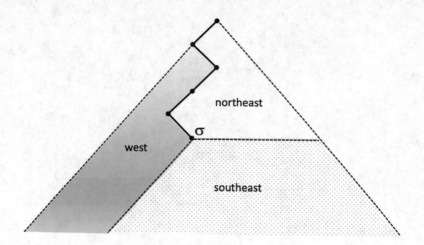

Preface

History

Classical computability theory (formerly known as recursive function theory) was pioneered by Turing, Post, Kleene and a few others in the late 1930s and throughout the 1940s. They developed intriguing concepts, some of which, such as reducibility and completeness, were adapted decades later for the study of concrete complexity. Also, Kleene's notation became the basis of the programming language Lisp. These pioneers asked tough questions and answered some of them. Post's Problem was formulated in 1944 and was deemed the holy grail of the subject for a dozen years; some feared that the discipline had played itself out, that all of the good ideas had already been discovered, that the grail would lie in eternal darkness (as grails are wont to do). Then two teenagers, Friedberg in the United States and Muchnik in the former Soviet Union, independently developed the "finite injury priority method," an elegant and robust algorithmic technique for constructing computably enumerable (c.e.) sets that satisfy certain "requirements." Not only did it solve Post's Problem, but it led to a flurry of related methods invented to construct c.e. sets subject to more complicated collections of requirements. The key idea was that a specific requirement could be "injured" only finitely often.

The very idea of an effective "infinite injury" method might seem impossible, but in the 1960s such methods were indeed invented and employed to prove startling results. Various models of infinite injury methods were developed; it's still not clear whether each one of them can simulate each of the others. In the 1970s an extremely complicated technique, called the "triple jump method" was developed. From a certain point of view, the triple jump method bears the same relationship to infinite injury that infinite injury bears to finite injury.

During the 1990s, the discipline again seemed played out (and so it may seem to this day). The latest methods had become so complicated that young logicians looked elsewhere for research topics. Classical computability theory had become a victim of its own success.

This Book

This book is focused on the algorithms used in the proofs, rather than on the results. They are presented in a unified notation and framework that will be more accessible not only to mathematicians but also to computer scientists. Scattered throughout the book are analogies to certain concepts that arise in computer science. Furthermore, the algorithms are presented in a pseudo-code notation that is typical of algorithms books such as [CLRS]. I am unaware of any other book or article in computability theory that describes algorithms in such pseudo-code.

As part of this focus on algorithms, a typical chapter in this book introduces a specific algorithmic technique. Such a chapter contains a single application, a theorem, to illustrate that technique. For example, Chap. 4 introduces the priority method, using a simple splitting theorem as the application. As another example, Chap. 9 introduces the length-of-agreement method, using a version of the Sacks Splitting Theorem as the application. When choosing the application, I strove to find the simplest, least general theorem to which the technique applies.

The motivation for this book is four-fold:

1. No book focusing on the algorithms for constructing c.e. sets currently exists. Also, some of the proofs here are worked out in greater detail and with more intuition than can be found anywhere else. This book lacks many of the standard topics found in books on computability (such as [Ro] or [So87]). For example, it does not mention reductions other than Turing reductions, or the arithmetical hierarchy, or many of the standard results of computability theory (such as Kleene's Recursion Theorem).
2. The algorithmic ideas developed for these abstract problems might find applications in more practical areas. For example, the priority tree—one of the central ideas of the book—involves multiple processes working on the same task, although with varying sets of assumptions or "guesses." At least one of those processes (and we cannot know a priori which it will be) will guess entirely correctly; that "true" node will accomplish the task. Is there an application for this idea amongst the more concrete problems?
3. In typical books on concrete algorithms, such as [CLRS], after discussing a few data structures, there is a series of chapters that could be permuted in just about any order without harming the book. It is not a defect in those books; it just appears to be the nature of the subject. On the other hand, the techniques that have been developed for constructing c.e. sets tend to build on each other. By focusing on these techniques, this book clarifies their relation to each other.
4. Beyond practicality, there is aesthetics. The violinist Isaac Stern said, "Every child deserves to know that there is beauty in the world." Every adult does, too. One who delves into length-of-agreement arguments and tree-based infinite injury arguments cannot help but marvel, with a wild surmise, like John Keats upon first looking into Chapman's Homer.[1]

---

[1] My apologies for waxing poetic. I'll try not to do it in the body of the text.

Included here are a plethora of exercises, most of which will differ greatly from those in other books on computability. Those books ask the reader to prove results, essentially using and perhaps combining the methods presented in the text. In this book, the exercises are mainly about the constructions themselves—e.g., why was this step done in this way? Or, consider the following modified version of our algorithm; would it work, too? The exercises are an integral part of the book—do them!

Acknowledgements

Carl Jockusch introduced me to recursion theory in a graduate course. A truly great teacher, he walked us through not only the material, but also (and perhaps more importantly) his own process of problem solving. Our homework assignments were not for the faint of heart; a typical assignment had five problems, with the instruction "DO ANY ONE" at the top of the page. Decades later, when I returned to the study of recursion theory (which had become known as computability theory), I peppered Prof. Jockusch with questions, which he graciously answered.

Denis Hirshfeldt and Peter Cholak also have been generous with their time, answering many of my questions.

Many students have contributed to this book, by their questions and ideas. Chief among them was Jacob Urick. He proof-read much of the book and made many, many suggestions for improving the notation, the organization of specific proofs, and the overall plan of the book. He also influenced the writing style. For example, he once told me, "Every time you explain something to me, you draw a picture. So, why not put them in your book?" Then I remembered how, in class, nearly every time Carl Jockusch explained something, he drew a picture. I took Jacob's advice, as you can see from the abundance of figures.

A Truth Universally Acknowledged

Consider this "result" stated in Chap. 13 of [AK]:

Theorem: *Recursion theory is very hard.*

As quipped in [AK], this result was "discovered independently by a number of other mathematicians."

It is "a truth universally acknowledged."[2]

Knowing this truth at the outset might discourage a student. My experience has been the opposite. When I'm struggling (for weeks, or for months) to understand a

---

[2] As Jane Austen would have said.

proof in computability theory, instead of giving up in frustration, I remind myself that this field is just, plain hard; I tell myself that my rate of progress is acceptable, perhaps even good. Occasionally, I look back and take stock of the terrain that I have traversed so far.

Enjoy the journey.

Columbus, USA   Kenneth J. Supowit

Contents

Chapter 1
Notation and Terms

1.1 Index of Notation and Terms

The following is a partial list of the notation and terms (some of which are informal) used in this book, categorized by the chapter in which they first appear. A few of these terms first appear in the exercises of the listed chapter.

Chapter 1

$\omega =_{def} \{0,\ 1,\ 2, \ldots\}$
λ

Chapter 2

$\aleph_0$
$\aleph_1$
(Cantor) diagonalization
requirement
witness

Chapter 3

Turing machine
$\Phi(k) \downarrow$
$\Phi(k) \uparrow$
$\Phi(k) = y$
lexicographic order
$|\sigma|$ (the length of string σ)
Φ_e (the eth Turing machine)
partial function
total function

© The Author(s), under exclusive license to Springer Nature Switzerland AG 2023
K. J. Supowit, *Algorithms for Constructing Computably Enumerable Sets*,
Computer Science Foundations and Applied Logic,
https://doi.org/10.1007/978-3-031-26904-2_1

$f(k) \downarrow$ (k is in the domain of the partial function f)
$f(k) \uparrow$ (k is not in the domain of the partial function f)
equality of two partial functions
Φ_e (the partial function computed by the eth Turing machine)
computable partial function
W_e
computably enumerable (c.e.) set
$\bar{A}$ (the complement of A)
computable set
c.e.n. set
computable enumeration
standard enumeration
$\langle x, y \rangle$
H (the "Halting Problem")

Chapter 4

how an infinite c.e. set is given as input
A_s ($= \{a_0, a_1, \ldots, a_s\}$)
stage
priority
the binary operation $\prec$ defined on requirements
            ($R \prec Q$ means that R has priority over Q)
$W_e[s]$ ($= \{w_{e,0}, w_{e,1}, \ldots, w_{e,s}\}$)
sparse set
simple set

Chapter 5

oracle Turing machine (OTM)
χ_A (the characteristic function of A)
Φ_e^B (the eth OTM with oracle B)
the computation $\Phi_e^B(k)$
$\Phi_e^B(k)$ (the output of the computation $\Phi_e^B(k)$)
Φ_e^B (the partial function computed by the eth OTM with oracle B)
φ_e^B (the usage function of Φ_e^B)
$\leq_T$
$\equiv_T$
$<_T$
(Turing) incomparable sets
prefix
proper prefix
the binary operation $\prec$ defined on strings rather than on requirements
            ($\tau \prec \sigma$ means that τ is a proper prefix of σ)
$\Phi_e^\sigma(n)$
$\sigma^\frown 0$ ($= \sigma$ appended with 0)

$\sigma^\frown 1$ ($= \sigma$ appended with 1)

A' (the jump of A)

Chapter 6

$A \upharpoonright n$ ($= A \cap \{0, 1, \ldots, n-1\}$)

$A \upharpoonright\upharpoonright n$ ($= A \cap \{0, 1, \ldots, n\}$)

the suffix $[s]$ (as in $\Phi_e^B(n)[s]$, or $\varphi_e^B(n)[s]$, for examples)

the computation $\Phi_e^B(n)[s]$ being permanent

Chapter 7

D_s (the constructed set D at the start of stage s)

requirement needing attention

requirement receiving attention

spoil a computation

injure a requirement

fresh witness

autoreducible set

Chapter 8

permission

$A(k)$ as a shorthand for $\chi_A(k)$, and accordingly A as a shorthand for χ_A

valid witness

Type 1, 2, or 3 witness

a witness holding up as Type 1, 2, or 3

Chapter 9

$\ell_{e,i}$ (the length of agreement)

$restraint_{e,i}$

the liminf of functions from ω to ω

Chapter 10

positive requirements

negative requirements

priority tree

Chapter 11

$A^{[e]}$

a row of a 0–1 matrix being full

piecewise trivial set

thick subset

tree of guesses, also known as a priority tree

$|\sigma|$

binary operations $\prec$, $\preceq$, $<_L$, $\vartriangleleft$, and $\trianglelefteq$ on tree nodes

TP_s

visiting a tree node

σ-stage

infinite path

TP (the true path)

$\sigma <_L TP$ (σ is to the left of TP)

$TP <_L \sigma$ (σ is to the right of TP)

$\sigma <_L TP_s$ (σ is to the left of TP_s)

$TP_s <_L \sigma$ (σ is to the right of TP_s)

nodes west, southeast, or northeast of σ

σ-believability

Φ_σ^B (a node-specific version of Φ_e^B)

ℓ_σ (a node-specific length of agreement)

$restraint_\sigma$

$Restraint_\sigma$

$\tilde{\ell}_\sigma$ (the maximum of ℓ_σ)

$\tilde{R}_\sigma$ (the maximum of $Restraint_\sigma$)

$pred_\sigma(s)$ for a σ-stage s

$A_s^{[e]}$

cofinite

I_σ (the injury set for node σ)

Chapter 12

$S \bigtriangleup T$ (the symmetric difference)

$\bar{\ell}_{e,r}$ (the length of disagreement)

unimodal function

Chapter 13

minimal pair of sets

positive witness for $P_{e,A}$ or $P_{e,B}$

negative witness for $P_{e,A}$ or $P_{e,B}$

$\Phi_e^A(k)[s] \downarrow = \Phi_e^B(k)[s]$

L_σ (the high water mark of ℓ_σ)

σ-expansionary stage

custody

Chapter 14

coding

n_σ (the witness list for node σ)

$m_\sigma(s)$

realized witness

unrealized witness

star witness

permanent star witness

undefining a witness
canceling a node
$inserted(\sigma, i)[s]$
$lastRealized(\sigma, i)[s]$
$\varphi(\sigma, i)[s]$
$entryIntoC(i)$
a fresh number (in this chapter, it must be odd)
a number unrealizes a witness
an inert node
a node performs an unrealization, insertion, or realization
a node takes the default
a witness is ready for insertion
window of opportunity

1.2 Defaults

We employ the following defaults, most of which are ubiquitous in the computability literature.

1. Lower case English letters (with the exception of f, g, and h, which are reserved for functions) refer to members of ω, which is defined as $\{0, 1, 2, \ldots\}$.
2. Starting in Chap. 3, *set* means a subset of ω, and *function* means a mapping from ω to ω (or, occasionally, from ω to $\{-1, 0, 1, 2, \ldots\}$). Upper case English letters refer to sets, unless specified otherwise (for example, whenever they refer to requirements rather than to sets, we say so).
3. Upper case Greek letters (typically Φ, and Ψ if we need a second one) refer to oracle Turing machines. Their corresponding lower case letters (φ and ψ) refer to their corresponding usage functions.
4. Typically, indices of machines or of requirements are e, i, or j. Inputs to machines are usually k, n, or p; in particular, witnesses are usually n. Stages of constructions are s, t, v, w, x, y, or z; in particular, x is typically a stage (which we cannot compute) after which certain finitely occurring events have ceased to occur.
5. The lower case Greek letters ρ, σ, τ, ξ and occasionally α, β, and γ refer to finite-length strings. Also, we use λ to denote the empty string. We use finite-length strings (if they are binary) to represent initial segments of characteristic functions, especially for oracles. Also, we use finite-length strings (whether or not they are binary) to represent tree nodes; in this context, λ represents the root of a tree. When considering a tree node σ, we often use ρ for a node to the left of σ, and τ for an ancestor of σ.

1.3   Notes About the Pseudo-code

To specify algorithms, we often use pseudo-code that resembles that of most text-books on algorithms, especially [CLRS].

Many of our **for** loops run forever, e.g.:

> **for** $s \leftarrow 0$ **to** ∞
>> statement
>> statement
>> statement

You won't see that in [CLRS]!

1.4   Miscellaneous Notes About the Text

Each chapter contains at most one theorem, which allows us to number each theorem by the chapter in which it appears (e.g., Theorem 11). Lemmas on the other hand, are numbered with both their chapter numbers and their positions among the lemmas in that chapter (e.g., Lemma 11.2); facts are numbered analogously.

Starting in Chap. 9, we often refer to the facts and lemmas without including the number of the current chapter. For example, in Chap. 14, we usually say "Fact 6" instead of its full name "Fact 14.6," and "Lemma 1" instead of its full name "Lemma 14.1."

Chapter 2
Set Theory, Requirements, Witnesses

This chapter is a very brief introduction to infinite cardinals. Our main interest here is in "requirements" and in "witnesses," which are key concepts throughout this book.

First, some definitions:

1. A *set* is a collection of elements.
2. The *Cartesian product* of sets A and B is

$$A \times B = \big\{ (a, b) : a \in A \text{ and } b \in B \big\}.$$

3. A *binary relation* on sets A and B is a subset of $A \times B$.
4. A *function* from set A to set B is a binary relation $f \subseteq A \times B$ such that

$$\big(\forall a \in A \big) \big(\exists b_1 \in B \big) \Big[(a, b_1) \in f \text{ and } (\forall b_2 \in B) \big[b_1 \neq b_2 \implies (a, b_2) \notin f \big] \Big].$$

The set A is the *domain*, and B is the *range* of f. We denote this situation by

$$f : A \to B$$

and we use the familiar infix notation $f(a) = b$ to denote $(a, b) \in f$.
5. A function $f : A \to B$ is

   (a) *one-to-one* if $\big(\forall a, a' \in A \big) \big[a \neq a' \implies f(a) \neq f(a') \big]$,
   (b) *onto* if $\big(\forall b \in B \big) \big(\exists a \in A \big) \big[f(a) = b \big]$,
   (c) a *one-to-one correspondence* if it is both one-to-one and onto.

The alternative, Latin-based terms are *injective*, *surjective*, and *bijective*, respectively.

© The Author(s), under exclusive license to Springer Nature Switzerland AG 2023
K. J. Supowit, *Algorithms for Constructing Computably Enumerable Sets*,
Computer Science Foundations and Applied Logic,
https://doi.org/10.1007/978-3-031-26904-2_2

6. A set S is *finite* if there is a one-to-one correspondence

$$f : S \to \{1, 2, \ldots, k\}$$

for some k.
7. A set S is *countably infinite* if there is a one-to-one correspondence $f : S \to \omega$. If so, then
$$S = \{ f^{-1}(0), \ f^{-1}(1), \ \ldots \}.$$

8. A set S is *countable* if it is either finite or countably infinite.

2.1 Diagonalization

If S is countably infinite then its cardinality is denoted by $\aleph_0$ (pronounced "aleph-null").

Here is a handy result for showing that certain sets are countable; it is proved in [Va]:

Lemma 2.1 *Let S be a set. If there exists a one-to-one (but not necessarily onto) function*
$$f : S \to \omega$$

then S is countable.

There would be no point in defining countability if all sets were countable. So here is one that is not ($\Re$ denotes the set of real numbers):

Theorem 2 *The set $S = (0, 1] = \{x \in \Re : 0 < x \le 1\}$ is not countable.*

Proof Assume for a contradiction that S were countable. Then there is a one-to-one correspondence $f : S \to \omega$. Let $a_i = f^{-1}(i)$ for each $i \in \omega$; thus

$$S = \{a_0, \ a_1, \ \ldots\}.$$

Let a_{ij} denote the jth digit in the decimal expansion of a_i, for each $i, j \in \omega$, where by convention no number has an infinite string of trailing 0's in its expansion (thus, we write $0.24999 \cdots$ instead of $0.25000 \cdots$). Let b be the real number $0.b_0 b_1 \cdots$, where

$$b_i = \begin{cases} 3, & \text{if } a_{ii} = 4 \\ 4, & \text{otherwise} \end{cases}$$

for each i. Thus $0 < b \le 1$ and so $b \in S$. However, for each i, b differs from a_i in the ith digit, and so $b \notin S$, a contradiction. $\qquad\qquad\square$

Fig. 2.1 Cantor
diagonalization

$$
\begin{aligned}
a_0 &= 0.\underline{a_{00}} \quad a_{01} \quad a_{02} \quad a_{03} \quad a_{04} \cdots \\
a_1 &= 0.a_{10} \quad \underline{a_{11}} \quad a_{12} \quad a_{13} \quad a_{14} \cdots \\
a_2 &= 0.a_{20} \quad a_{21} \quad \underline{a_{22}} \quad a_{23} \quad a_{24} \cdots \\
a_3 &= 0.a_{30} \quad a_{31} \quad a_{32} \quad \underline{a_{33}} \quad a_{34} \cdots \\
a_4 &= 0.a_{40} \quad a_{41} \quad a_{42} \quad a_{43} \quad \underline{a_{44}} \cdots \\
&\vdots
\end{aligned}
$$

The argument in the proof of Theorem 2 is called "diagonalization," because of the key role of the diagonal elements, which are underlined in Fig. 2.1. It is also known as "Cantor diagonalization," after its discoverer, the pioneer of set theory, Georg Cantor.

The cardinality of the set $(0, 1]$ is denoted by $\aleph_1$. Here is the list of cardinalities that we have seen so far:

$$0, 1, 2, \ldots, \aleph_0, \aleph_1.$$

Are there others? If so, how many are there? What does it mean for one infinite cardinal to be larger than another? These questions are (partially) answered in the next section.

2.2 Infinitely Many Infinite Cardinals

The cardinality of set A is *less than or equal* to the cardinality of set B, denoted $|A| \leq |B|$, if there is a one-to-one function from A to B. Their cardinalities are *equal*, denoted $|A| = |B|$, if there is a one-to-one correspondence from A to B; otherwise we write $|A| \neq |B|$. This generalizes our concept of what it means for two finite sets to be equal-sized; if you gave a little kid a bunch of crayons and pencils, and asked him whether there's the same number of each, he might just pair them up and then see whether anything is left over. The cardinality of A is *less than* that of B, denote $|A| < |B|$, if $|A| \leq |B|$ but $|A| \neq |B|$.

Lemma 2.2 *Let A and B be sets. Then exactly one of the following three conditions holds:*

I. $|A| < |B|$,
II. $|B| < |A|$,
III. $|A| = |B|$.

Furthermore, if $|A| \leq |B|$ and $|B| \leq |A|$ then $|A| = |B|$.

Although Lemma 2.2 may look trivial, proving it see to require some effort (see [Va] for a nice proof).

The *power set* of a set A, denoted $POW(A)$ is the set of all subsets of A. E.g.,

$$POW(\{a, b, c\}) = \{ \{\}, \{a\}, \{b\}, \{c\}, \{a, b\}, \{a, c\}, \{b, c\}, \{a, b, c\} \}.$$

Lemma 2.3 *Let A be a set. Then $|A| < |POW(A)|$.*

Proof First note that $|A| \leq |POW(A)|$, because the function $f : A \to POW(A)$ defined by

$$f(a) = \{a\}$$

is one-to-one.

Next we prove $|A| \neq |POW(A)|$ by showing that there is no one-to-one correspondence from A to $POW(A)$. To do this, let $g : A \to POW(A)$; we will show that g is not onto. Let

$$S = \{ a \in A : a \notin g(a) \}.$$

Thus $S \in POW(A)$. Assume for a contradiction that S were an image under g, that is, $S = g(a_0)$ for some $a_0 \in A$.

Case 1: $a_0 \in S$.
  That is, $a_0 \in \{ a \in A : a \notin g(a) \}$. Thus $a_0 \notin g(a_0) = S$, a contradiction.

Case 2: $a_0 \notin S$.
  That is, $a_0 \notin \{ a \in A : a \notin g(a) \}$. Thus $a_0 \in g(a_0) = S$, a contradiction.

Thus g is not onto, and g was an arbitrarily chosen function from A to $POW(A)$. □

Corollary $|\omega| < |POW(\omega)| < |POW(POW(\omega))| < \cdots$.

Thus, there are infinitely many infinite cardinals.

2.3 What's New in This Chapter?

Take another look at the proof of Theorem 2. The idea is to construct a number $b \in \{a_0, a_1, \ldots\}$ that satisfies, for each i, the *requirement*

$$R_i : b \neq a_i.$$

If each of these infinitely many requirements is satisfied, then we have a contradiction. We construct b so that the diagonal element a_{ii} is a *witness* to the requirement R_i. This extra terminology—requirements, witnesses—sheds no additional light on Cantor's simple proof. However, as this book progresses, we shall see more complicated types of requirements, and beautiful techniques for obtaining witnesses to them.

2.4 Exercises

1. Give examples to show that

    (a) the intersection of two countably infinite sets can be either finite or countably infinite,

    (b) the intersection of two uncountable sets can be finite, countably infinite, or uncountable.

2. How many subsets are there of ω? How many of them are finite?
3. Let S be the union of countably many countable sets. Must S be countable?
4. A real number is *rational* if it can be expressed as the ratio of two integers. Let $\mathbb{Q}$ denote the set of rational numbers. Show that $\mathbb{Q}$ is countable.
5. Suppose that we tried to use diagonalization to show that $\mathbb{Q}$ is uncountable, mimicking the proof of Theorem 2. Where would the argument fail?
6. Let S be a set of non-overlapping discs, each having positive area (that is, none of these discs is merely one point), and each contained entirely within the unit square $[0, 1] \times [0, 1]$. Prove that S is countable.

Chapter 3
Computable and c.e. Sets

Starting here, and for the remainder of this book, the term *set* refers to a subset of ω (unless specified otherwise). In this chapter we define what it means for a set to be "computable" and "computably enumerable." To do so, we first (briefly) describe a formal model of computation, called the "Turing machine."

3.1 Turing Machines

Informally, a Turing machine Φ is depicted in Fig. 3.1.

The machine Φ has a two-way infinite tape, and a read/write head whose movements are determined by a "controller."[1] The cells of the tape are numbers by integers. In Fig. 3.1, the read/write head is scanning cell -2, which contains the symbol 1. The machine moves in discrete time steps. At all times, the controller is in exactly one "state." The set of possible states is

$$Q = \left\{ q_{start}, \, q_{halt}, \, q_1, \, q_2, \, \ldots, \, q_r \right\} \tag{3.1}$$

for some $r \in \omega$. Thus, Q is finite. At all times, each tape cell contains either 0 or 1 or B (which stands for "blank"). At the start of each time step, depending on the current state and on the symbol currently being scanned by the read/write head, either 0 or 1 or B is written into the cell being scanned (call it cell i), and then the read/write head moves left (that is, to cell $i - 1$) or right (that is, to cell $i + 1$).

A Turing machine is specified as a pair (r, δ), where

$$r = |Q| - 2$$

---

[1] Think of the controller as a finite automaton, if you are familiar with that type of machine. Otherwise, never mind; the description here is self-contained.

© The Author(s), under exclusive license to Springer Nature Switzerland AG 2023
K. J. Supowit, *Algorithms for Constructing Computably Enumerable Sets*,
Computer Science Foundations and Applied Logic,
https://doi.org/10.1007/978-3-031-26904-2_3

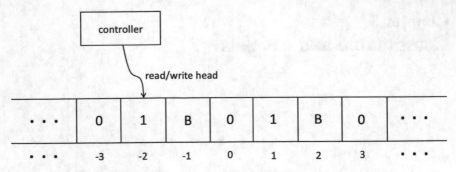

Fig. 3.1 A Turing machine (without an oracle)

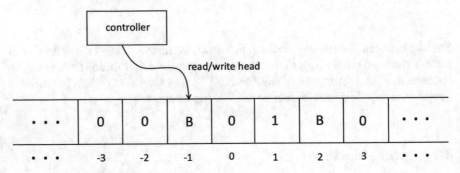

Fig. 3.2 The machine of Fig. 3.1, one step later

as in (3.1), and

$$\delta : \left(Q - \{q_{halt}\} \right) \times \{0, 1, B\} \ \rightarrow \ Q \times \{0, 1, B\} \times \{left, right\}$$

is the *transition function*. Informally, δ tells the machine how to move. For example, if

$$\delta(q_8, 1) = (q_5, 0, right)$$

and at the start of step s the machine is in state q_8 and looks like Fig. 3.1, then at the start of step $s + 1$ the machine would be in state q_5 and look like Fig. 3.2.

Note that δ depends only on the current state and the symbol being scanned; it does *not* depend on the number of the cell being scanned.

As input, Φ takes a member of ω, represented in binary without leading 0s.[2] When the machine is turned on (that is, at time 0), the read/write head is scanning cell 1, and the binary string representing the input k begins at cell 1 and ends at cell $\lceil \log_2(k + 1) \rceil$; all other cells contain B at time 0. Thus, if the machine looks like Fig. 3.3 at time 0, then its input is 11.

---

[2] The empty string λ represents 0.

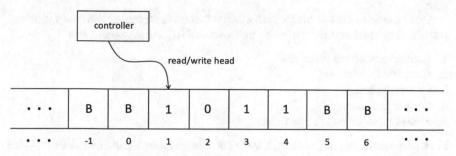

Fig. 3.3 A Turing machine at time 0 with input 11

The machine Φ begins at time 0 in state q_{start}. Suppose that Φ is started on input k. If Φ eventually enters state q_{halt} then Φ halts and we say that Φ *accepts* k; we denote this situation by $\Phi(k) \downarrow$. Otherwise, we say that Φ *rejects* k, which we denote by $\Phi(k) \uparrow$. If and when Φ with input k halts, its output is defined as the number y whose binary representation (which may contain leading 0's) appears in cells $1, 2, \ldots, m-1$, where

$$m = \min\{c : c \geq 1 \text{ and cell } c \text{ contains symbol } B\},$$

and we write

$$\Phi(k) = y.$$

Such an m must exist because Φ begins with only finitely many non-blanks on its tape, and can change at most one blank to 0 or 1 during each step; hence the number of non-blanks is finite when Φ halts. Therefore, for each k,

$$\Phi(k) \downarrow \implies (\exists y)\big[\Phi(k) = y\big].$$

Numerous variations of the Turing machine have been defined and found convenient for various purposes. As examples:

1. The cells could contain symbols from any finite "alphabet," not just $\{0, 1, B\}$.
2. The number of tapes could be greater than one. It could even be countably infinite.
3. The transition function δ could be "non-deterministic," which necessitates a broader definition of what it means for the machine to accept.

Details of these and other "loaded" versions of the Turing machine are discussed in textbooks such as [Ma] or [Si], where they are shown to be no more powerful than our stripped-down model. That is, a stripped-down machine can simulate the operation of a loaded machine in a step-by-step fashion.

Other models or notations for computation that (at first glance) seem very different from Turing machines have been defined and studied. Among them are:

1. Kleene's recursive functions,
2. Church's λ-calculus,
3. C++ programs,
4. Fortran programs,
5. certain types of cellular automata.

Each such model of computation, it turns out, can simulate Turing machines and vice versa. These experiences have lent credence to the following widely-held idea:

Church's Thesis[3]: Anything that can reasonably be described as discrete, automatic computation can be simulated by a Turing machine.

Throughout this book we freely and implicitly use Church's Thesis. That is, whenever we describe an algorithm in pseudo-code, or prove or assume that an algorithm with a certain input-output relation exists, we implicitly assume that the algorithm can be simulated by a Turing machine. Think of Turing machine notation as just another programming language. So, why do we bother to introduce such a primitive language? Why not instead base this entire book on the set of Python programs? We could, but Turing machines help us visualize and reason about oracles, an essential concept starting in Chap. 5.

3.2 Computably Enumerable, Computable, c.e.n

Each Turing machine has a finite description, which can be encoded as a binary string, because the transition function δ can be specified by a finite number of bits. These finite-length binary strings can be put into lexicographic order.[4] Therefore the number of Turing machines is $\aleph_0$. We name them as Φ_0, Φ_1, ..., according to lexicographic order.

---

[3] Also known as the "Church-Turing Thesis.".

[4] If α and β are strings, then α precedes β *lexicographically* if either $|\alpha| < |\beta|$ or ($|\alpha| = |\beta|$ and α precedes β alphabetically), where $|\gamma|$ denotes the length of string γ. Thus, the entire set of finite-length binary strings in lexicograph order is:

$$\lambda, \ 0, \ 1, \ 00, \ 01, \ 10, \ 11, \ 000, \ 001, \ \ldots$$

The following algorithm, given e, outputs a description of Φ_e:

ALGORITHM 3.1

> $count \leftarrow -1$.
> **for** $i \leftarrow 0$ **to** ∞
> > **if** the ith binary string in lexicographic order is a
> > > syntactically correct description of a Turing machine Φ
> > > $count \leftarrow count + 1$.
> > > **if** $count = e$
> > > > output this description of Φ.
> > > > halt.

More definitions:

1. A *partial function* is a function

$$f : D \to \omega$$

for some $D \subseteq \omega$. The set D is the *domain* of f. The partial function f is a *(total) function* if $D = \omega$. We write $f(k) \downarrow$ if $k \in D$; otherwise we write $f(k) \uparrow$.

2. If f and g are partial functions, then

$$f = g$$

means that for each k,

$$\left(f(k) \uparrow \text{ and } g(k) \uparrow \right) \quad \text{or} \quad \left(f(k) \downarrow \text{ and } g(k) \downarrow \text{ and } f(k) = g(k) \right).$$

3. Turing machine Φ_e defines a partial function, which we denote by Φ_e. Thus, the symbol Φ_e refers both to a Turing machine and to the partial function that it defines; the context will disambiguate it.

4. A partial function $f : \omega \to \omega$ is *computable* if

$$(\exists e)\left[\Phi_e = f \right].$$

5. Let

$$W_e =_{def} \left\{ k : \Phi_e(k) \downarrow \right\}.$$

In other words, W_e is the domain of the partial function Φ_e.

6. A set A is *computably enumerable (c.e.)* if

$$(\exists e)\left[A = W_e \right].$$

Thus, W_0, W_1, ... constitute the c.e. sets.

7. The *complement* of A is

$$\bar{A} =_{def} \omega - A.$$

8. If A and $\bar{A}$ are both c.e. then A is *computable* (as is $\bar{A}$).
9. If A is c.e. but not computable then A is *c.e.n.*
10. A *computable enumeration* of a set A is a sequence of finite sets

$$A_0 \subseteq A_1 \subseteq \cdots$$

such that

(i) $\bigcup_i A_i = A$,
(ii) there is an algorithm that, given i, outputs the set A_i.

11. A *standard enumeration* of A is a sequence of distinct numbers $a_0, a_1, \ldots$ such that

$$A = \{a_0, a_1, \ldots\},$$

and that the total function $f : \omega \to A$ defined by $f(i) = a_i$ is computable.

Lemma 3.1 *A set has a standard enumeration* $\Longleftrightarrow$ *it has a computable enumeration.*

The proof of Lemma 3.1 is left as Exercise 6.

Lemma 3.2 *A set has a standard enumeration* $\Longleftrightarrow$ *it is c.e.*

Proof Fix A.

First, suppose that A has a standard enumeration $\{a_0, a_1, \ldots\}$. The following algorithm is given k as input:

ALGORITHM 3.2
    for $i \leftarrow 0$ **to** ∞
      if $a_i = k$
        accept.

Thus,

$$A = \{k : \text{Algorithm 3.2 accepts } k\}.$$

By Church's Thesis, there is a Turing machine that implements Algorithm 3.2, and so A is c.e.[5]

To show the converse, assume that A is c.e.; that is, there exists e such that $A = W_e$. Consider the following algorithm, which takes no input:

---

[5] This is our first and last explicit use of Church's Thesis, but (as already pointed out) we will often use it implicitly.

ALGORITHM 3.3

    for $s \leftarrow 1$ **to** ∞
        for $k \leftarrow 0$ **to** s
            if (Φ_e halts on input k within s steps) and (k has not yet been output)
            output(k).

Because Algorithm 3.3 outputs no number more than once, and because

$$(\forall k)[\, k \in A \iff \text{Algorithm 3.3 outputs } k \,],$$

Algorithm 3.3 outputs a standard enumeration of A.   □

    Lemmas 3.1 and 3.2 together imply that a set is c.e. if and only if it has a computable enumeration; this explains the etymology of "computably enumerable."

    What is the etymology of "computable?" Suppose A is computable, that is, A and $\bar{A}$ are both c.e. Then there exist e and e' such that $A = W_e$ and $\bar{A} = W_{e'}$. Then the following algorithm, given k, ascertains whether $k \in A$:

ALGORITHM 3.4

    for $s \leftarrow 1$ **to** ∞
        if Φ_e halts on input k within s steps
           output("k is in A") and halt.
        if $\Phi_{e'}$ halts on input k within s steps
           output("k is not in A") and halt.

    Algorithm 3.4 is guaranteed to halt; hence the problem of determining membership in A is "computable."

3.3 An Example of a c.e.n. Set

Let $< i, j >$ denote the pairing function defined in Appendix A. Let

$$H =_{def} \left\{ \langle e, k \rangle : \ \Phi_e(k) \downarrow \right\}.$$

The set H is called "the halting problem," the problem being, given e and k, to ascertain whether $\langle e, k \rangle \in H$. We will show that H is c.e.n.

Fact 3.1 H *is c.e.*

Proof Consider the following algorithm, that is given m as input:

ALGORITHM 3.5

> $e, k \leftarrow$ the unique numbers such that $m = \langle e, k \rangle$.
> Use Algorithm 3.1 to find a description of Φ_e.
> **for** $s \leftarrow 1$ **to** ∞
> > **if** Φ_e halts on input k within s steps
> > > halt.

Algorithm 3.5 halts on input m if and only if $m \in H$. Hence H is c.e. □

Fact 3.2 *H is not computable.*

Proof Assume that H were computable. Then the total function

$$f(\langle e, k \rangle) = \begin{cases} 1, & \text{if } \Phi_e(k) \downarrow \\ 0, & \text{if } \Phi_e(k) \uparrow \end{cases}$$

is computable. Note that f is a function of one variable, call it m, but we think of it as a function of two variables e and k such that $m = \langle e, k \rangle$.

The partial function

$$g(i) = \begin{cases} 5, & \text{if } f(\langle i, i \rangle) = 0 \\ \uparrow, & \text{if } f(\langle i, i \rangle) = 1 \end{cases}$$

is computable; let j be such that $\Phi_j = g$. Then

$$g(j) \downarrow \implies f(\langle j, j \rangle) = 0 \implies \Phi_j(j) \uparrow \implies g(j) \uparrow$$

and

$$g(j) \uparrow \implies f(\langle j, j \rangle) = 1 \implies \Phi_j(j) \downarrow \implies g(j) \downarrow.$$

Thus, whether or not $g(j) \downarrow$, we have a contradiction. □

Facts 3.1 and 3.2 together imply:

Lemma 3.3 *H is c.e.n.*

In subsequent chapters, many results begin "Let A be c.e.n. Then ..." By Lemma 3.3, such results are not vacuously true.

3.4   What's New in This Chapter?

This chapter introduces Turing machines, Church's Thesis, and the definitions of computable, c.e., and c.e.n. sets.

Chapter 5 will introduce Turing machines augmented with an "oracle;" upon those enhanced machines all subsequent chapters will be built.

3.5   Afternotes

More thorough introductions to Turing machines and evidence for Church's Thesis can be found in books on computability, such as [Ma], which is highly intuitive and has many good examples and exercises, or [Ro] and [Si], which are more concise.

Kleene's recursive function notation is described and studied in [Ro].

In the literature, during the 1990s or so, the adjective "computable" replaced the older term "recursive," just as "computably enumerable (c.e.)" replaced "recursively enumerable (r.e.)." "Decidable" is another synonym for "computable," as is "Turing-recognizable" for "c.e."

Some authors write "c.e.a" (abbreviating "c.e. above 0") to denote what we are calling "c.e.n."

3.6   Exercises

1. For each of the following statements, say whether it is true for all A, and prove your answer:

   (a) A is computable $\iff$ $\bar{A}$ is computable.
   (b) A is c.e. $\iff$ $\bar{A}$ is c.e.
   (c) At least one of $\{A, \bar{A}\}$ is c.e.
   (d) A is computable $\iff$ A has a standard enumeration $a_0 < a_1 < \cdots$
       (in other words, A is computable if and only if its elements can be computably enumerated in increasing order).

2. Let A be c.e.n.

   (a) Does A have a non-c.e. subset?
   (b) Does A have an infinite, computable subset?

3. Let A be non-computable, and let

$$B_0 = \{k \in A : k \text{ is even}\}$$

and

$$B_1 = \{k \in A : k \text{ is odd}\}.$$

(a) Must at least one of $\{B_0, \ B_1\}$ be computable?

(b) Must at least one of $\{B_0, \ B_1\}$ be non-computable?

4. Let A be c.e.n. Might $\bar{A}$ be c.e.?

5. Let A be c.e. Prove that the set

$$\{e : A = W_e\}$$

is infinite.

6. Prove Lemma 3.1.

7. If we change line 4 of Algorithm 3.4 to

if $\Phi_{e'}$ halts on input k within 2^s steps

would Algorithm 3.4 still be guaranteed to halt?

8. We proved Lemma 3.3 mainly to show that there exists a c.e.n. set. Alternatively, could the existence of a c.e.n. set be shown by a cardinality argument?

9. Let $a_0, \ a_1, \ \ldots$ be a standard enumeration of an infinite set A. Let

$$T = \left\{t : \ (\forall s > t)[a_s > a_t]\right\}.$$

Prove that T is infinite.

Chapter 4
Priorities (A Splitting Theorem)

In the beginning, there was diagonalization. Take another look at Sect. 2.3, which introduces the ideas of requirements and witnesses.

In this chapter, we construct sets B_0 and B_1 that satisfy certain requirements. Initially, both B_0 and B_1 are empty. Our algorithm works in infinitely many "stages." During each stage, it adds just one number either to B_0 and to B_1. The work during each stage is performed in a finite amount of time, which ensures that both B_0 and B_1 are c.e. The requirements (which are infinite in number) might interfere with each other; to maintain law and order, we assign priorities to the requirements, and each one is eventually satisfied.

This technique is called a "priority argument." We will see more priority arguments, in Chaps. 7–9, 11–14.

4.1 A Priority Argument

Theorem 4 *Let A be c.e.n. Then A can be partitioned into c.e.n. sets B_0 and B_1.*

Proof We describe an algorithm that, given a c.e.n. set A, partitions it into c.e.n. sets B_0 and B_1. What does it mean to be "given" an infinite c.e. set? We are actually given a number j such that $W_j = A$. We could use j to output a standard enumeration $a_0, a_1, \ldots$ of A as follows:

> **for** $s \leftarrow 1$ **to** ∞
>   **for** $k \leftarrow 0$ **to** s
>   **if** k has not been output yet
>   **if** Φ_j halts on input k within s steps
>   output(k).

© The Author(s), under exclusive license to Springer Nature Switzerland AG 2023
K. J. Supowit, *Algorithms for Constructing Computably Enumerable Sets,*
Computer Science Foundations and Applied Logic,
https://doi.org/10.1007/978-3-031-26904-2_4

Here we did not specify how to produce a description of Φ_j, given j; for that, see Algorithm 3.1.[1]

For each s, let

$$A_s =_{def} \{a_0, a_1, \ldots, a_s\}.$$

We proceed in stages: during each stage s, we put a_s into B_0 or B_1 (but not into both), thereby ensuring that B_0 and B_1 partition A. Each stage will be computable in a finite amount of time, and so both B_0 and B_1 are c.e. We also need to ensure that both B_0 and B_1 are non-computable. How will we do that?

A simple approach would be to assign a_0, a_2, a_4, ... to B_0, and a_1, a_3, a_5, ... to B_1. This would ensure that either B_0 or B_1 is non-computable (because otherwise the set $A = B_0 \cup B_1$ would be computable), but there's no reason why both of them must be.

Instead of this simple alternation of assignments, we use a more subtle approach. In particular, we construct B_0 and B_1 so as to satisfy the requirements

$$R_{e,i} : W_e \neq \overline{B_i}$$

(i.e., W_e and B_i are not complementary) for each $e \in \omega$ and $i \in \{0, 1\}$. Collectively, the requirements $R_{e,0}$ ensure that $\overline{B_0}$ is not c.e., and hence that B_0 is not computable. Likewise, the requirements $R_{e,1}$ collectively ensure that $\overline{B_1}$ is not c.e., and hence that B_1 is not computable. During each stage, we try to meet (i.e., satisfy) a requirement that is not yet met. There are infinitely many unmet requirements; we choose to meet the one of strongest "priority," where the requirements are prioritized as follows:

$$R_{0,0} \prec R_{0,1} \prec R_{1,0} \prec R_{1,1} \prec R_{2,0} \prec R_{2,1} \prec \cdots.$$

Two sets are not complementary if and only if there is an n in both of them or in neither. Therefore we meet requirement $R_{e,i}$ by producing an n that is in both of W_e and B_i, or in neither; such an n is a witness for $R_{e,i}$. We have some control over B_i, because we can assign elements of A either to B_i or to B_{1-i}. However, we have no control over W_e. Therefore, because W_e might not be computable, we have no way to determine whether a given n, not yet enumerated in W_e, eventually will be.

Our algorithm is as follows, where $w_{e,0}$, $w_{e,1}$, ... is a standard enumeration of W_e, and

$$W_e[s] =_{def} \{w_{e,0}, w_{e,1}, \ldots, w_{e,s}\}. \tag{4.1}$$

What it means for requirement $R_{e,i}$ to *act* is defined by the comment just before line 8 of Algorithm 4.1.

---

[1] In the remainder of this book, we won't specify how we produce a description of Φ_e given e, or how we are "given" a c.e. set.

ALGORITHM 4.1

```
1  B₀ ← ∅.
2  B₁ ← ∅.
3  for s ← 0 to ∞
       // This is stage s.
       // Assign aₛ either to B₀ or to B₁.
4      assigned ← FALSE.
5      for e ← 0 to s
6          for i ← 0 to 1
7              if (not assigned) and aₛ ∈ Wₑ[s] and Wₑ[s] ∩ Bᵢ = ∅
                   // Requirement Rₑ,ᵢ acts.
8                  Put aₛ into Bᵢ.     // This action ensures that Wₑ ∩ Bᵢ ≠ ∅.
9                  assigned ← TRUE.
10     if not assigned
11         Put aₛ into B₀.     // Or put aₛ into B₁, it doesn't matter.
```

Note that for each $i \in \{0, 1\}$,

$$\bar{A} \subseteq \overline{B_i} \tag{4.2}$$

(the Venn diagram in Fig. 4.1 may be helpful for seeing this).

We need to verify the algorithm; that is, we need to show that each requirement is met. Fix e. We will show that the requirement $R_{e,0}$ is met (the proof that $R_{e,1}$ is met is analogous, substituting $R_{e,1}$ for $R_{e,0}$, and B_1 for B_0).

Assume for a contradiction that $R_{e,0}$ were not met; in other words, assume that

$$W_e = \overline{B_0}. \tag{4.3}$$

Let

$$A' = \left\{ a : (\exists s)\left[a = a_s \in W_e[s] \right] \right\}.$$

Fig. 4.1 B_0 and B_1 partition A

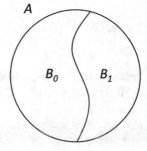

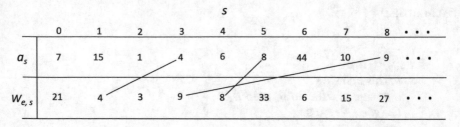

Fig. 4.2 $\{4, 8, 9\} \subseteq A'$

Informally, A' consists of all numbers that are enumerated in W_e before, or at the same time as, being enumerated in A. In Fig. 4.2, such numbers are indicated by line segments. Note that

$$A' \subseteq A \cap W_e$$

and that

$$\left(\forall t\right)\left[\, w_{e,t} \in A \implies w_{e,t} \in A_t \cup A'\,\right] \tag{4.4}$$

(see Exercise 2).

Case 1: A' is infinite.

Only finitely many requirements have stronger priority than that of $R_{e,0}$, and no requirement acts more than once. Hence there exists a stage x after which none of those stronger priority requirements acts. Because A' is infinite, there exists $s > x$ such that $a_s \in A'$.

Because $s > x$, when the condition in line 7 of Algorithm 4.1 is evaluated so as to decide whether $R_{e,0}$ will act during stage s, it must be that

$$assigned = \text{FALSE}.$$

Because $a_s \in A'$, we have

$$a_s \in W_e[s].$$

Therefore

$$W_e[s] \cap B_0 \neq \emptyset, \tag{4.5}$$

because otherwise $R_{e,0}$ would act during stage s, contradicting (4.3). However, (4.5) itself contradicts (4.3).

Case 2: A' is finite.

In other words, there are only finitely many line segments in Fig. 4.2.

Proposition: $\bar{A}$ is c.e.

Proof Consider the following algorithm.

ALGORITHM 4.2
1 **for** $t \leftarrow 0$ **to** ∞
2   **if** $w_{e,t} \notin A_t \cup A'$
3   output $w_{e,t}$.

Fix p.

Case 2.1: $p \in \bar{A}$.

Then $p \in \overline{B_0} = W_e$ (by (4.2) and (4.3)) and so $p = w_{e,t}$ for some t. Because $p = w_{e,t} \notin A$ and $A_t \cup A' \subseteq A$, we have

$$w_{e,t} \notin A_t \cup A',$$

and therefore Algorithm 4.2 outputs $w_{e,t} = p$.

Case 2.2: $p \notin \bar{A}$.

If $p \notin W_e$ then Algorithm 4.2 does not output p. So assume $p \in W_e$; then there exists t such that

$$p = w_{e,t}.$$

Therefore, because $p \in A$, we have

$$w_{e,t} \in A_t \cup A',$$

by (4.4). Hence the condition in line 2 of Algorithm 4.2 evaluates to false, and so again Algorithm 4.2 does not output p.

Thus, Algorithm 4.2 outputs p if and only if $p \in \bar{A}$; in other words, it enumerates $\bar{A}$. The condition in line 2 of Algorithm 4.2 can be evaluated in a finite amount of time, because

$$A_t \cup A'$$

is finite (because we are in Case 2). Hence $\bar{A}$ is c.e.

QED Proposition

Because A is c.e., the Proposition implies that A is computable, contrary to the assumption that A is c.e.n. This concludes Case 2.

In summary, in either Case 1 or Case 2, the assumption of (4.3) leads to a contradiction. Therefore $R_{e,0}$ is met.

QED Theorem 4

We have just seen a simple priority argument. The essence of its simplicity is that once a requirement $R_{e,i}$ acts by putting a witness into B_i, that witness remains credible forever after, and so $R_{e,i}$ stays met. This will not be true in Chap. 7 and beyond, where a requirement R might act but then later get "injured," that is, its witness might lose its credibility, and so we would need to find a new witness for R. Chapter 7 introduces the "finite injury priority method," where each requirement is injured only finitely often. Chapter 10 introduces the beautiful "infinite injury priority method," where a specific requirement may be injured infinitely many times (and yet somehow it all works out).

Hang on to your hat.

4.2   What's New in This Chapter?

We just saw our first priority argument, which involved the following new ideas:

1. Priorities among requirements.
2. The dynamic choosing of witnesses. That is, witnesses are chosen during the algorithm, rather than being pre-defined as they were in the diagonalization argument of Theorem 2.
3. The definition of a certain stage, which we typically call x, after which certain finitely occurring events never occur. We used this technique in Case 1 of the proof of Theorem 4. It is used in all priority arguments (as far as I know). We argue that such an x exists, even though we need not, and perhaps cannot, compute it.[2]

4.3   Afternotes

The splitting theorem presented here is a weaker version of the Friedberg Splitting Theorem ([Fr58]).

---

[2] Hence our priority-based proofs are "non-constructive." This may be a bit confusing, because algorithms in this discipline are traditionally called "constructions." Thus, such a construction is part of a non-constructive proof.

4.4 Exercises

1. Suppose that we replace Algorithm 4.1 by the following:

> $B_0 \leftarrow \emptyset$.
> $B_1 \leftarrow \emptyset$.
> **for** $s \leftarrow 0$ **to** ∞
> > // Assign a_s either to B_0 or to B_1.
> > *assigned* $\leftarrow$ FALSE.
> > $t \leftarrow 0$.
> > **while** not *assigned*
> > > **for** $e \leftarrow 0$ **to** t
> > > > **for** $i \leftarrow 0$ **to** 1
> > > > > **if** (not *assigned*) and $\left(a_s \in W_e[t]\right)$ and $\left(W_e[t] \cap B_i = \emptyset\right)$
> > > > > // Requirement $R_{e,i}$ *acts*.
> > > > > Put a_s into B_i.
> > > > > *assigned* $\leftarrow$ TRUE.
> > $t \leftarrow t + 1$.

   Would the **while** loop halt for each s? Would each requirement $R_{e,i}$ be met?
2. Prove (4.4).
3. Let A be c.e.n. Then for each k, we can partition A into c.e.n. sets $B_0, B_1, \ldots, B_k$ by applying Theorem 4, a total of k times. Describe a way to partition A into infinitely many c.e.n. sets $B_0, B_1, \ldots$
4. An infinite set S is *sparse* if for each i,

$$\left| S \cap \{2^i, \, 2^i + 1, \, \ldots, \, 2^{i+1} - 1\} \right| \le 1.$$

   Show that if A is c.e.n., then A can be partitioned into c.e.n. sets B_0 and B_1 such that B_0 is sparse.
5. For each $e \in \omega$ and $i \in \{0, 1\}$, let

$$S_{e,i} : W_e \ne B_i$$

   (compare $S_{e,i}$ to $R_{e,i}$), where B_0 and B_1 are the sets constructed by Algorithm 4.1. Is it possible that $S_{e,i}$ is met for each $e \in \omega$ and $i \in \{0, 1\}$?
6. Suppose that we delete the words

$$\text{"and } W_e[s] \cap B_i = \emptyset\text{"}$$

   from line 7 of Algorithm 4.1. Explain why the proof would be invalid.

7. (from [Po]) A set B is *simple* if

   (i) B is c.e.,
   (ii) $\overline{B}$ is infinite,
   (iii) $\overline{B}$ contains no infinite, c.e. set.

Use a priority argument to prove that a simple set exists.

Hint: Use the requirements

$$R_e : W_e \text{ is infinite} \implies W_e \cap B \neq \emptyset.$$

Another hint: Your algorithm should construct B so that, for each e,

$$\big| B \cap \{0, 1, \ldots, 2e\} \big| \leq e$$

(this will ensure that $\overline{B}$ is infinite).

Chapter 5
Reductions, Comparability (Kleene-Post Theorem)

5.1 Oracle Turing Machines

An oracle Turing machine (OTM) is an ordinary Turing machine that has been enhanced with an "oracle tape," which has a read-only head, as is illustrated in Fig. 5.1. The oracle tape contains an infinite sequence of 0s and 1s. We interpret that sequence as the *characteristic function* of some set B, defined as

$$\chi_B(k) = \begin{cases} 1, & \text{if } k \in B \\ 0, & \text{otherwise.} \end{cases}$$

Thus, in Fig. 5.1, we see that 2 and 4 are in B, whereas 0, 1, and 3 are not. Thus, an OTM is given an input k on its read/write tape as depicted in Fig. 3.3, as well as χ_B on its read-only tape. The set B is called the *oracle* of the computation.

The transition function is now of the form:

$$\delta : (Q - \{q_{halt}\}) \times \{0, 1, B\} \times \{0, 1\} \rightarrow Q \times \{0, 1, B\} \times \{left, right\} \times \{left, right\}.$$

For example, if

$$\delta(q_6, \text{ B}, 0) = (q_{14}, 1, \textit{left}, \textit{right})$$

and at the start of step s the machine is in state q_6 and looks like Fig. 5.1, then at the start of step $s + 1$ the machine would be in state q_{14} and look like Fig. 5.2.

Think of the oracle tape as an infinite lookup table. An OTM with oracle B is just like a plain Turing machine, except that whenever it wants to, it can look up in the table whether a particular number is in B.

© The Author(s), under exclusive license to Springer Nature Switzerland AG 2023   31
K. J. Supowit, *Algorithms for Constructing Computably Enumerable Sets*,
Computer Science Foundations and Applied Logic,
https://doi.org/10.1007/978-3-031-26904-2_5

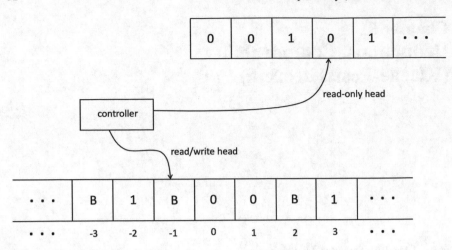

Fig. 5.1 An oracle turing machine

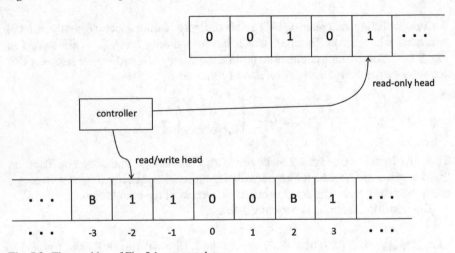

Fig. 5.2 The machine of Fig. 5.1, one step later

Just like a plain Turing machine, an OTM has a finite description, because even though a particular oracle is an infinite binary string, the OTM itself is specified merely by its transition function δ. Therefore, like the plain Turing machines, the oracle Turing machines can be ordered. Let Φ_e^B denote the eth OTM with oracle B.

The *computation* $\Phi_e^B(k)$ denotes the behavior of the eth OTM with input k and oracle B. If that computation halts, then we write $\Phi_e^B(k) \downarrow$ (otherwise we write $\Phi_e^B(k) \uparrow$). If it halts with output y, then we write $\Phi_e^B(k) = y$. We write $\Phi_e^B(k) \neq y$ if either

$$\Phi_e^B(k) \uparrow \quad \text{or} \quad \Phi_e^B(k) = \text{some number other than } y.$$

Thus, Φ_e^B defines a partial function. In analogy to Φ_e, the notation Φ_e^B refers both to an OTM and to the partial function that it defines; the context will disambiguate it.

Often, we are interested in the *usage function*:

$$\varphi_e^B(k) =_{def} \begin{cases} \text{the rightmost position of the oracle head} \\ \quad \text{during the computation } \Phi_e^B(k), & \text{if } \Phi_e^B(k)\downarrow \\ -1, & \text{otherwise.} \end{cases}$$

Informally, $\varphi_e^B(k)$ is the largest number about which the oracle is consulted during the computation $\Phi_e^B(k)$ (if it halts). Note that φ is a lower-case Greek phi, whereas Φ is an upper-case Greek phi.

5.2 Turing Reductions

We write $A \leq_T B$ if

$$(\exists e)(\forall k)\big[\, \chi_A(k) = \Phi_e^B(k) \,\big],$$

which is often abbreviated as

$$(\exists e)\big[\, \chi_A = \Phi_e^B \,\big],$$

and we say that A *(Turing-)reduces to* B. Informally, $A \leq_T B$ means that B is at least as hard (computationally) as A.

Lemma 5.1 contains some simple properties of Turing reductions; we leave its proof as Exercise 1.

Lemma 5.1 *For all A, B, and C,*

I. $A \leq_T A$ *(reflexivity)*
II. $A \leq_T \bar{A}$
III. $(A \leq_T B)$ *and* $(B \leq_T C) \implies (A \leq_T C)$ *(transitivity)*
IV. $(A \leq_T B)$ *and* B *is computable* $\implies A$ *is computable.*

If $A \leq_T B$ and $B \leq_T A$ then we write

$$A \equiv_T B$$

and say that A and B are *Turing equivalent*. We write

$$A <_T B$$

if $A \leq_T B$ but $B \not\leq_T A$. Sets A and B are *(Turing) incomparable* if $A \not\leq_T B$ and $B \not\leq_T A$.

Three questions arise, in analogy with those that arose in the theory of infinite cardinals[1]:

1. Is there an infinite sequence of sets A_0, A_1, ... such that $A_0 <_T A_1 <_T \cdots$?
2. Are there Turing incomparable sets?
3. Is there a set A such that $\emptyset <_T A <_T H$? Recall that H is the halting problem, defined in Chap. 3. We phrased this question in terms of the empty set $\emptyset$, but could have done so in terms of any computable set, because all computable sets are Turing equivalent to each other.

The answer to all three questions is "yes." Exercise 10(a) answers question (5.1). Theorem 5 answers questions (2) and (3).

5.3   The Theorem

Theorem 5   (Kleene-Post) *There exist incomparable sets A and B below H, that is,*

$$A \not\leq_T B,$$
$$B \not\leq_T A,$$
$$A <_T H,$$
$$B <_T H.$$

Proof   We construct incomparable sets A and B in stages. In particular, we define a sequence of finite-length binary strings

$$\alpha_0 \prec \alpha_1 \prec \alpha_2 \prec \cdots$$

where $\alpha_j \prec \alpha_{j+1}$ means that a_j is a proper prefix of a_{j+1}. A *prefix* of a string $c_0 c_1 \cdots c_k$ is defined as a string $c_0 c_1 \cdots c_j$ for some $j \leq k$; the prefix is *proper* if $j < k$. Informally, $\alpha_j \prec \alpha_{j+1}$ means that a_{j+1} is an extension of a_j. The use of the binary operator "$\prec$" on strings is unrelated to its use on requirements; the context will disambiguate between them.

The set A is defined by its characteristic function

$$\chi_A = \lim_{j \to \infty} \alpha_j,$$

as is illustrated in Fig. 5.3.

---

[1] In Chap. 2, we used the power set operation to obtain an infinite sequence of ever larger sets. By Lemma 2.2, all pairs of sets are comparable in cardinality. The question of whether there exists a set strictly larger than the integers but smaller than the reals is the famous "continuum hypothesis;" it's a long story (see [Coh]).

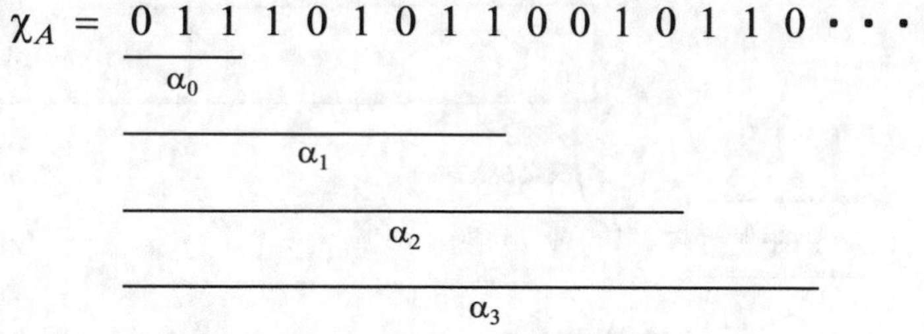

$$\chi_A = \underline{0\ 1\ 1}\ 1\ 0\ 1\ 0\ 1\ 1\ 0\ 0\ 1\ 0\ 1\ 1\ 0\ \cdots$$

$$\alpha_0$$

$$\alpha_1$$

$$\alpha_2$$

$$\alpha_3$$

Fig. 5.3 Defining a set A by a sequence of finite-length binary strings

Analogously, we define

$$\beta_0 \prec \beta_1 \prec \beta_2 \prec \cdots,$$

and B by its characteristic function

$$\chi_B = \lim_{j \to \infty} \beta_j.$$

Sets A and B will meet the requirements (for all e):

$$R_{2e} : \chi_A \neq \Phi_e^B$$
$$R_{2e+1} : \chi_B \neq \Phi_e^A.$$

Collectively, the even-numbered requirements guarantee that $A \not\leq_T B$, and the odd-numbered requirements guarantee that $B \not\leq_T A$. Requirement R_{2e} can be written as

$$(\exists n)[\,\chi_A(n) \neq \Phi_e^B(n)\,];$$

such an n is a witness for R_{2e}. Algorithm 5.1, which constructs A and B, comes up with a witness for each R_{2e}, and one for each R_{2e+1} (i.e., an n such that $\chi_B(n) \neq \Phi_e^A(n)$).

Here is some more notation used in the pseudo-code for Algorithm 5.1, where σ is a binary string of finite length:

1. $\sigma \frown 0$ denotes σ appended with a 0, and $\sigma \frown 1$ denotes σ appended with a 1.
2. $\Phi_e^\sigma(n) \downarrow$ means that the eth OTM, with the first $|\sigma|$ bits on its oracle tape constituting σ, and with input n, halts, and during that computation the oracle head is never to the right of cell $|\sigma|$. Figure 5.4 illustrates such a machine at time 0; the word "irrelevant" is based on the assumption that $\Phi_e^\sigma(n) \downarrow$ (Fact 5.1 below clarifies this).

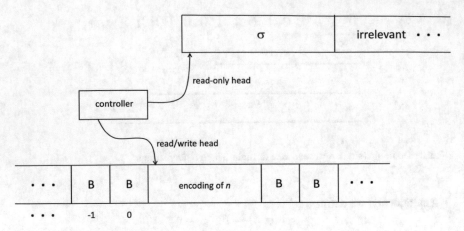

Fig. 5.4 The start of a halting computation, with σ as an oracle

If $\Phi_e^\sigma(n) \downarrow$ then the machine must halt with some output y; we denote this situation by

$$\Phi_e^\sigma(n) = y.$$

In Algorithm 5.1, our only interest is whether

$$\Phi_e^\sigma(n) = 1.$$

The key idea behind Algorithm 5.1 is:

Fact 5.1 *For all e, n, σ, and τ,*

$$\left(\sigma \prec \tau \ \text{ and } \ \Phi_e^\sigma(n) \downarrow \right) \implies \left(\Phi_e^\sigma(n) = \Phi_e^\tau(n) \right).$$

Proof During the computation $\Phi_e^\tau(n)$, the oracle head never moves past the right boundary of σ. $\qquad\Box$

ALGORITHM 5.1

1   $\alpha_0 \leftarrow \lambda$.   // Recall that λ denotes the empty string.
2   $\beta_0 \leftarrow \lambda$.
3   **for** $e \leftarrow 0$ **to** ∞
        // Define α_{2e+1} and β_{2e+1} so as to meet R_{2e}.
4       $n \leftarrow |\alpha_{2e}|$.   // Thus, n is the least number whose
                            // membership in A is yet to be specified.
5   **if** $(\exists \sigma)\big[\, \beta_{2e} \prec \sigma$ and $\Phi_e^\sigma(n) = 1 \,\big]$
6           $\alpha_{2e+1} \leftarrow \alpha_{2e} \frown 0$.   // Thus, $n \notin A$.
7           $\beta_{2e+1} \leftarrow \sigma$.   // Thus, $\Phi_e^B(n) = 1$.
        else
8           $\alpha_{2e+1} \leftarrow \alpha_{2e} \frown 1$.   // Thus, $n \in A$.
9           $\beta_{2e+1} \leftarrow \beta_{2e} \frown 0$.   // Or $\beta_{2e} \frown 1$, it doesn't matter.

        // Define α_{2e+2} and β_{2e+2} so as to meet R_{2e+1}.
10      $n \leftarrow |\beta_{2e+1}|$.   // Thus, n is the least number whose
                            // membership in B is yet to be specified.
11   **if** $(\exists \sigma)\big[\, \alpha_{2e+1} \prec \sigma$ and $\Phi_e^\sigma(n) = 1 \,\big]$
12          $\beta_{2e+2} \leftarrow \beta_{2e+1} \frown 0$.   // Thus, $n \notin B$.
13          $\alpha_{2e+2} \leftarrow \sigma$.   // Thus, $\Phi_e^A(n) = 1$.
        else
14          $\beta_{2e+2} \leftarrow \beta_{2e+1} \frown 1$.   // Thus, $n \in B$.
15          $\alpha_{2e+2} \leftarrow \alpha_{2e+1} \frown 0$.   // Or $\alpha_{2e+1} \frown 1$, it doesn't matter.

We need to verify the algorithm, that is, to prove that all the requirements are met. Consider requirement R_{2e}. Suppose the condition in line 5 of Algorithm 5.1 evaluates to true, that is,

$$(\exists \sigma)\big[\, \beta_{2e} \prec \sigma \text{ and } \Phi_e^\sigma(n) = 1 \,\big].$$

Then

$$\chi_A(n) = 0 \neq 1 = \Phi_e^\sigma(n) = \Phi_e^B(n),$$

by Fact 5.1. On the other hand, if

$$\neg(\exists \sigma)\big[\, \beta_{2e} \prec \sigma \text{ and } \Phi_e^\sigma(n) = 1 \,\big]$$

then

$$\chi_A(n) = 1 \neq \Phi_e^B(n),$$

because either

$$\Phi_e^B(n) \uparrow \quad \text{or} \quad \Phi_e^B(n) = \text{some number other than 1.}$$

In either case n is a witness for R_{2e}, and hence R_{2e} is met. In the latter case, we extend β_{2e} with either 0 or 1 (see line 9 of Algorithm 5.1). In that case, why do we bother to extend β_{2e} at all? We do it to ensure that χ_B is a total function.

The proof that R_{2e+1} is met is analogous.

Suppose that we had an oracle for H. Then we could ascertain whether

$$(\exists \sigma) \left[\beta_{2e} \prec \sigma \text{ and } \Phi_e^{\sigma}(n) = 1 \right] \tag{5.1}$$

and, if so, produce such a σ; likewise we could ascertain whether

$$(\exists \sigma) \left[\alpha_{2e+1} \prec \sigma \text{ and } \Phi_e^{\sigma}(n) = 1 \right] \tag{5.2}$$

and, if so, produce such a σ (see Exercise 5). Therefore

$$A \leq_T H \quad \text{and} \quad B \leq_T H,$$

which implies

$$A <_T H \quad \text{and} \quad B <_T H$$

(see Exercise 6).

$$QED \text{ } Theorem \text{ } 5$$

If A and B are incomparable then neither is computable, because a computable set reduces to every set. Thus, we have now seen two ways to guarantee that a constructed set C is non-computable:

1. ensure that $W_e \neq \overline{C}$ for each e, as we did in the proof of Theorem 4,
2. construct a set D such that $C \nleq D$, as we did in the proof of Theorem 5.

The incomparable sets A and B constructed by Algorithm 5.1 do not appear to be c.e., because while determining (*without* the help of an oracle for H) whether (5.1) or whether (5.2) is true, the algorithm could search forever for the extension σ. In the vernacular, Algorithm 5.1 might "go out to lunch and not come back." In Chap. 7, we will see a more subtle algorithm that constructs a pair of incomparable sets in such a way that each stage is guaranteed to run in a finite amount of time, and so those two sets will be c.e.

5.4   What's New in This Chapter?

1. Oracle Turing machines.
2. Usage functions.
3. Turing reductions.

4. Defining a set by an infinite sequence of finite prefixes of its characteristic function.
5. Protecting a requirement from future injury. In Theorem 5, we were able to guarantee such protection; however, in Chap. 7 and beyond, we will try to avoid injuries but will not always succeed.

5.5 Afternotes

In this book, the only reductions that we study are Turing reductions. However, many other types of reductions have been defined and studied (see [Ro]). Turing reductions seem the most natural, because $A \leq_T B$ corresponds to the idea of algorithm A using algorithm B as a subroutine.

The practice of using upper-case Greek letters to denote OTM computations and corresponding lower-case Greek letters to denote their usage functions has become common in the literature. For example, a paper might refer to OTM's Φ and Ψ and their corresponding usage functions φ and ψ. This notation may seem strange at first, but it grows on you.

5.6 Exercises

1. Prove Lemma 5.1.
2. Is $\leq_T$ commutative?
3. If $A \leq_T B$ and B is c.e., must A be c.e.?
4. Let B be computable. Prove that for each A,

$$A \leq_T B \iff A \text{ is computable}$$

   (thus, computable oracles are useless for our purposes).
5. Describe an algorithm that uses an oracle for H to ascertain whether (5.1) is true and, if so, to produce such a σ. Do likewise for (5.2).
6. Let A, B, and C be sets such that A and B are incomparable, $A \leq_T C$, and $B \leq_T C$. Prove that $A <_T C$ and $B <_T C$.
7. Find the flaw in the following argument: Sets A and B constructed by Algorithm 5.1 are each enumerated in ascending order. Therefore, by Exercise 1(d) of Chap. 3, they are both computable.
8. In Fig. 5.3,

$$\alpha_0 = 011$$
$$\alpha_1 = 011101011$$
$$\alpha_2 = 0111010110010$$
$$\alpha_3 = 0111010110010110$$

However, this situation is impossible if the α strings are constructed by Algorithm 5.1. Why?

9. The *jump* of A is defined as

$$A' = \{ e : \Phi_e^A(e) \downarrow \}.$$

Prove the following, for all A and B:

(a) A is computable $\implies A' \equiv_T H$
(b) $A <_T A'$
(c) $A \leq_T B \implies A' \leq_T B'$

10. (a) Prove that there is an infinite sequence

$$A_0 <_T A_1 <_T \cdots .$$

(b) Prove that if $A_0 <_T A_1 <_T \ldots$ then

$$(\exists B)(\forall i)[A_i <_T B].$$

Chapter 6
The Permanence Lemma

This chapter introduces some handy notation, and one little—but quite useful—lemma about computations with c.e. oracles.

6.1 Notation

The following three pieces of notation help to simplify expressions, and are ubiquitous in the literature. At first glance they may seem unimportant, but their value will become apparent as we delve into more complicated arguments.

1. $A \upharpoonright\upharpoonright k =_{def} A \cap \{0, 1, \ldots k\}$.
2. $A \upharpoonright k =_{def} A \cap \{0, 1, \ldots k - 1\}$.
3. The suffix "[s]" appended to an expression means that everything that can be modified by a stage s, is so modified.
   As examples:

   (a) $\Phi_e^B(k)[s]$ denotes the result of the computation $\Phi_e^{B_s}(k)$ after s steps (note that B_s is the oracle here rather than B).
   (b) $\varphi_e^B(k)[s]$ denotes the usage of the computation $\Phi_e^{B_s}(k)$ after s steps. Note that

   $$\left(\forall e', C, k, z \right) \left[\varphi_{e'}^C(k)[z] \leq z \right],$$

   because the oracle head starts at cell 0 of the oracle tape and moves at most one position during each step. In the extreme case, it moves to the right on each step, resulting in

   $$\varphi_{e'}^C(k)[z] = z.$$

   (c) $(C \cap D)[s] = C_s \cap D_s$.

6.2   The Lemma

Suppose that D is a c.e. set, with a computable enumeration

$$D_0 \subseteq D_1 \subseteq \cdots$$

Fix k and e, and suppose that $\Phi_e^D(k) \downarrow$. Must it be that

$$\Phi_e^D(k) = \lim_s \Phi_e^D(k)[s] \; ?$$

Must it be that

$$\varphi_e^D(k) = \lim_s \varphi_e^D(k)[s] \; ?$$

The answer to both questions is "yes;" this follows from Lemma 6.1, which we call the "Permanence Lemma."

   These answers might not be obvious; one might imagine a sequence

$$t_0 < t_1 < \cdots$$

such that

1. $\Phi_e^D(k)[t_0] \downarrow$
2. some $d_1 \leq \varphi_e^D(k)[t_0]$ enters D after stage t_0 (that is, $d_1 \in D - D_{t_0}$), and causes $\Phi_e^D(k)[t_1] \uparrow$
3. $\Phi_e^D(k)[t_2] \downarrow$
4. some $d_2 \leq \varphi_e^D(k)[t_2]$ enters D after stage t_2, and causes $\Phi_e^D(k)[t_3] \uparrow$

and so forth. Thus, $\lim_s \Phi_e^D(k)[s]$ would not exist. In this scenario, is it possible that

$$\Phi_e^D(k) \downarrow ?$$

The answer to the question is "no," which follows from the Permanence Lemma.

   We say that a computation $\Phi_e^D(k)[s]$ is *permanent* if

$$\Phi_e^D(k)[s] \downarrow \quad \text{and} \quad D_s \restriction u = D \restriction u \, ,$$

where $u = \varphi_e^D(k)[s]$. Informally, this says that no number in $D - D_s$ is small enough to affect the computation $\Phi_e^D(k)[s]$. Note that if the computation $\Phi_\sigma^D(k)[s]$ is permanent then for each $t \geq s$,

$$\Phi_e^D(k)[t] = \Phi_e^D(k)[s] \quad \text{and} \quad \varphi_e^D(k)[t] = \varphi_e^D(k)[s] \, .$$

For that matter, all other aspects of the computation $\Phi_\sigma^D(k)[t]$ are the same as for $\Phi_\sigma^D(k)[s]$; for example, if the read/write head is positioned at cell 37 exactly 52 times during the computation $\Phi_\sigma^D(k)[s]$ then it will do likewise during the computation $\Phi_\sigma^D(k)[t]$.

Lemma 6.1 (Permanence Lemma) *Let $D_0 \subseteq D_1 \subseteq \cdots$ be a computable enumeration of D. Suppose $\Phi_e^D(k) \downarrow$. Then there exists an s such that the computation $\Phi_e^D(k)[s]$ is permanent.*

Proof Let x be such that the computation $\Phi_e^D(k)$ halts in fewer than x steps. Let $s > x$ be such that

$$D_s \upharpoonright \varphi_e^D(k) = D \upharpoonright \varphi_e^D(k). \tag{6.1}$$

The computation $\Phi_e^{D_s}(k)$ is identical to the computation $\Phi_e^D(k)$; hence it, too, halts in fewer than x steps, and therefore it halts in fewer than s steps. Hence

$$\varphi_e^D(k) = \varphi_e^D(k)[s].$$

Therefore, by (6.1),

$$D_s \upharpoonright \varphi_e^D(k)[s] = D \upharpoonright \varphi_e^D(k)[s],$$

and so the computation $\Phi_e^D(k)[s]$ is permanent. $\qquad\square$

6.3 Afternotes

The suffix "$[s]$" notation was introduced in [La79].

6.4 Exercises

1. Suppose that the computation $\Phi_e^D(k)[s]$ is permanent. Must it be that

$$\Phi_e^D(k) = \Phi_e^D(k)[s]?$$

2. In the proof of the Permanence Lemma, must the computation $\Phi_e^D(k)[x]$ be permanent?

3. Suppose $d_0, d_1, \ldots$ is a standard enumeration of D, and e and k are such that

$$(\exists y)(\forall z \geq y)\big[\, \Phi_e^D(k)[z] = 5 \, \big].$$

Must $\Phi_e^D(k) = 5$?

Chapter 7
Finite Injury (Friedberg-Muchnik Theorem)

In this chapter, the Kleene-Post construction of Chap. 5 is enhanced with the idea of priorities introduced in Chap. 4, along with a key new idea: injury.

7.1 The Theorem

Theorem 7 (Friedberg-Muchnik) *There exist incomparable, c.e. sets.*

Proof In Algorithm 7.1, initially $A = B = \emptyset$. During each stage s, a finite number of elements might be added to A, and a finite number to B. Let A_s and B_s denote the sets A and B, respectively, at the start of stage s.[1] Then

$$A_0 \subseteq A_1 \subseteq \cdots$$

and

$$B_0 \subseteq B_1 \subseteq \cdots.$$

The set A is defined as $\bigcup_s A_s$, and B as $\bigcup_s B_s$. This is more flexible than the Kleene-Post construction (Algorithm 5.1), because there the elements of A were inserted in increasing order (likewise for B), whereas here that might not happen.

The requirements are identical to those used in the Kleene-Post Theorem, namely:

$$R_{2e} : \chi_A \neq \Phi_e^B$$
$$R_{2e+1} : \chi_B \neq \Phi_e^A.$$

---

[1] We use this notation, namely D_s to denote a constructed set D at the start of stage s, from here until the end of this book.

© The Author(s), under exclusive license to Springer Nature Switzerland AG 2023
K. J. Supowit, *Algorithms for Constructing Computably Enumerable Sets*,
Computer Science Foundations and Applied Logic,
https://doi.org/10.1007/978-3-031-26904-2_7

Algorithm 5.1 met the requirements $R_0, R_1, \ldots$ in that order. It could get stuck while working on a particular requirement, and so the constructed sets were not necessarily c.e. Algorithm 7.1 meets the requirements in a more complicated order.

We prioritize the requirements as follows:

$$R_0 \prec R_1 \prec \cdots .$$

The strategy for meeting a single requirement R_{2e} is to attach to it a potential witness n not yet in A. If we eventually reach a stage s such that

$$n \notin A_s \text{ and } \Phi_e^B(n)[s] = 0,$$

then we say that R_{2e} *needs attention* at stage s.

Analogously, R_{2e+1} *needs attention* at stage s if

$$n \notin B_s \text{ and } \Phi_e^A(n)[s] = 0,$$

for its witness n.

At the start of a given stage s, many requirements might need attention. We attend to the one of strongest priority. That is, we pick the least j such that R_j needs attention. Suppose that this j is even (otherwise we would swap the roles of A and B); so $j = 2e$ for some e. We put n_j, which is the witness for R_j, into A, and we say that R_j *receives attention*. Then we try to restrain all numbers less than or equal to $\varphi_e^B(n_j)[s]$ from entering B. If we succeed in restraining all such numbers from entering B during or after stage s, then we would have

$$B_s \restriction \varphi_e^B(n_j)[s] = B \restriction \varphi_e^B(n_j)[s]. \tag{7.1}$$

If (7.1) is true then the computation $\Phi_e^B(n_j)[s]$ would be "protected" or "preserved;" in particular, we would have

$$\Phi_e^B(n_j)[s] = \Phi_e^B(n_j) = 0$$

and so R_j would be met.

On the other hand, if R_j never needs attention at stage s or beyond, that is,

$$(\forall t \geq s)\big[\Phi_e^B(n_j)[t] \neq 0 \big],$$

then we would never put n_j into A. In this case, we would have

$$n_j \notin A \text{ and } \Phi_e^B(n_j) \neq 0$$

and so, again, R_j would be met. Note that $\Phi_e^B(n_j)[t] \neq 0$ means either

$$\Phi_e^B(n_j)[t]\uparrow \text{ or } \big(\Phi_e^B(n_j)[t] = \text{ some number other than } 0 \big).$$

This simple construction might fail, because we might not be able to keep certain numbers less than or equal to $\varphi_e^B(n_j)[s]$ from entering B. Such a number might enter B during a stage $t \geq s$ at the behest of a stronger priority requirement, and thereby "spoil" the computation $\Phi_e^B(n_j)[s]$ (that is, "injure" R_j). The term *injury*, though widely used, is a bit misleading. The entrance of such a small number into B during stage t may or may not invalidate the witness n_j (that is, it may or may not cause $\Phi_e^B(n_j) \neq 0$). So, rather than saying that witness n_j is "injured," it is more accurate to say that it has "lost credibility." However, we need not ascertain whether R_j has actually become injured; rather, we take a cautious approach and assume that if it might have been injured, it was. It's like finding an open box of cereal on the shelf of a grocery store; its contents may or may not have been "injured," but we play it safe by reaching for a different box.

Thus, our witness may lose credibility.[2] If it does, then we choose a *fresh* witness for R_j and keep watching until R_j again needs attention (which might never happen). By "fresh," we mean that the number is greater than every witness chosen so far, and greater than the usage (that is, the ϕ value) of every computation performed so far.

These ideas are clarified in the pseudo-code for Algorithm 7.1. It calls a subroutine *Initialize(k)* that assigns a fresh witness to R_k (that is, it assigns a fresh number to the variable n_k). What it means for a requirement R_j to *act* is defined by the comment just before line 8 of Algorithm 7.1.

ALGORITHM 7.1

```
1    A ← ∅.
2    B ← ∅.
3    n₀ ← 0.   // This is equivalent here to Initialize(0).
4    for s ← 1 to ∞
5        Initialize(s).
6        for j ← 0 to s
7            if Rⱼ needs attention
                 // Rⱼ acts.
8                Put nⱼ into A (if j is even; otherwise put it into B).
9                Initialize(k) for each k such that j < k ≤ s.
```

A finite amount of work is performed during each stage of Algorithm 7.1 (whereas Algorithm 5.1 could get stuck in one stage). Thus, the sets A and B constructed by this algorithm are c.e.

---

[2] In other words, our witness may be tampered with.

The theorem is implied by the following lemma:

Lemma 7.1 *For each j, R_j acts only finitely often, and is met.*

Proof We proceed by strong induction on j.

For the basis, consider $j = 0$. The witness for R_0, namely the variable n_0, receives the value 0 by line 3 of Algorithm 7.1, and never changes, because there is no stronger priority requirement than R_0. If R_0 ever needs attention, then it acts and can never subsequently be injured, and so

$$0 \in A \text{ and } \Phi_e^B(0) = 0.$$

Otherwise (R_0 never needs attention)

$$0 \notin A \text{ and } \Phi_e^B(0) \neq 0$$

($\Phi_e^B(0) \neq 0$ follows from the converse of the Permanence Lemma). In either case, R_0 is met.

For the inductive step, fix $j \geq 1$. Suppose that $j = 2e$ for some e; otherwise $j = 2e + 1$ and the roles of A and B would be swapped. Assume the claim for each $i < j$. Then the variable n_j has a final value; call it n. If R_j ever needs attention after n_j receives n, then it acts and can never subsequently be injured, and so

$$n \in A \text{ and } \Phi_e^B(n) = 0.$$

Otherwise (R_j never needs attention after n_j receives n)

$$n \notin A \text{ and } \Phi_e^B(n) \neq 0$$

(as in the basis, $\Phi_e^B(0) \neq 0$ here follows from the converse of the Permanence Lemma). In either case, R_j is met.

QED Lemma 7.1

QED Theorem 7

7.2 What's New in This Chapter?

1. A priority argument with injuries.
2. A witness "needing attention."
3. A "fresh" witness.

7.3 Afternotes

The Kleene-Post algorithm gave us a set A such that $\emptyset <_T A <_T H$. This led to the question:

$$\text{Is there a c.e. set } A \text{ such that } \emptyset <_T A <_T H?$$

This question, known as Post's Problem, was posed in 1944, and was considered the central question in recursive function theory (which is now called "computability theory") until it was answered in the affirmative (it's a corollary of Theorem 7) in 1956, independently by two teenagers—an American named Friedberg [Fr57] and a Russian named Muchnik [Mu]. Friedberg and Muchnik solved the problem by inventing essentially the same technique, the "finite-injury priority method," which we saw in this chapter. The method turned out to have many more uses, both in computability theory and in other branches of logic.

In [Ku], the Friedberg-Muchnik Theorem was proved without using a priority argument.

7.4 Exercises

1. Why is it necessary that distinct requirements have distinct witnesses?
2. (a) Give an upper bound on the number of times that R_0 can act.
   (b) Do the same thing for R_1.
   (c) Do the same thing for R_j, for all $j \geq 1$.
   (d) Modify Algorithm 7.1 so as to obtain a smaller upper bound than was found in part (c).
3. (a) Prove that for each s, at the start of stage s,

$$n_0 < n_1 < \cdots < n_{s-1}$$

   (note that for each $j \geq s$, at the start of stage s, variable n_j has never received a value).
   (b) For each j, the value of the variable n_j changes only finitely often. Let $\tilde{n}_j$ denote the final value of n_j, and let

$$\tilde{N} = \{\tilde{n}_j : j \in \omega\}.$$

   Is $\tilde{N}$ c.e.?
4. Modify Algorithm 7.1 so that not only are the constructed sets A and B incomparable and c.e., but also

   (a) A and B are sparse (as defined in Exercise 4 of Chap. 4).
   (b) $A \subseteq C$ and $B \subseteq C$, where C is a given infinite, c.e. set.

(c) $A \subseteq B$.
(d) all three of the above.

5. Prove that there is an infinite family of c.e. sets whose members are pairwise incomparable.
6. (from [Tr]) A set A is *autoreducible* if $(\exists e)(\forall k)$

$$k \in A \implies \Phi_e^{A-\{k\}}(k) = 1$$

and

$$k \notin A \implies \Phi_e^{A-\{k\}}(k) = 0.$$

In other words, the eth oracle Turing machine determines whether a given k is in A by asking its oracle questions of the form "Is $m \in A$?" for various m's other than k.
Describe an algorithm that constructs a c.e. set A that is *not* autoreducible.
Hint: Use a finite-injury priority argument. Define requirements R_e that collectively ensure that A is not autoreducible. Then define what it means for R_e to need attention at stage s. Finally, write pseudo-code for your algorithm.
7. Describe an algorithm that constructs a pair of incomparable, non-autoreducible, c.e. sets.
8. Let $t : \omega \to \omega$ be a computable total function.

(a) Show that there exists a computable total function

$$f : \omega \to \{0, 1\}$$

such that every Turing machine that computes f requires more than $t(n)$ steps to compute $f(n)$ for at least one value of n. This says, for example, that there is a computable function, with range $\{0, 1\}$, that cannot be computed in worst-case time less than 2^{2^n}.
Hint: diagonalize.
(b) Strengthen the claim of part (a) by showing that it remains true with "at least one value" replaced by "infinitely many values."
(c) Show that it remains true with "at least one value" replaced by "all but finitely many values."
Hint: use a priority argument.

Chapter 8
Permitting (Friedberg-Muchnik Below C Theorem)

Suppose that we are given a c.e.n. set C, and we wish to construct a c.e. set A *below* C (that is, $A \leq_T C$), while A satisfies certain other properties as well. Permitting is a way to do this.

The idea is that a number may enter A only when C permits it. There are various types of permitting; the simplest is based on the following lemma.

8.1 The Lemma

Lemma 8.1 (Permitting Lemma) *Let* $A_0 \subseteq A_1 \subseteq \cdots$ *and* $C_0 \subseteq C_1 \subseteq \cdots$ *be computable enumerations of A and C, respectively. If*

$$(\forall s, n)\left[n \in A_{s+1} - A_s \implies C_s \upharpoonright\upharpoonright n \neq C \upharpoonright\upharpoonright n \right] \tag{8.1}$$

then $A \leq_T C$.

Proof Assume

$$(\forall s, n)\left[C_s \upharpoonright\upharpoonright n = C \upharpoonright\upharpoonright n \implies n \notin A_{s+1} - A_s \right], \tag{8.2}$$

which is equivalent to (8.1). We will show that $A \leq_T C$. Assume that we have an oracle for C. Fix n. To determine whether $n \in A$, we first compute (with the help of the oracle) the finite set $C \upharpoonright\upharpoonright n$. Then we find a stage r such that

$$C_r \upharpoonright\upharpoonright n = C \upharpoonright\upharpoonright n.$$

© The Author(s), under exclusive license to Springer Nature Switzerland AG 2023
K. J. Supowit, *Algorithms for Constructing Computably Enumerable Sets*,
Computer Science Foundations and Applied Logic,
https://doi.org/10.1007/978-3-031-26904-2_8

Note that

$$(\forall s \geq r)\left[\, C_s \upharpoonright\! n = C \upharpoonright\! n \,\right].$$

Therefore, by (8.2),

$$(\forall s \geq r)\left[\, n \notin A_{s+1} - A_s \,\right],$$

and so

$$n \notin A - A_r.$$

Hence, to determine whether $n \in A$, it suffices to determine whether $n \in A_r$.

8.2   The Theorem

Recall that the Friedberg-Muchnik algorithm produced an incomparable pair of c.e. sets. We enhance that algorithm with permitting, to obtain the following stronger result:

Theorem 8   (Friedberg-Muchnik below C) *Let C be c.e.n. There exists c.e. A and B such that*

$$A \nleq_T B$$
$$B \nleq_T A$$
$$A <_T C$$
$$B <_T C.$$

Theorem 8 generalizes both Theorem 5 (the Kleene-Post Theorem) and Theorem 7 (the Friedberg-Muchnick Theorem). In Theorem 8 the constructed sets A and B must be c.e., unlike in Theorem 5; also, in Theorem 8 the given set C is an arbitrary c.e.n. set rather than specifically H as in Theorem 5.

Proof   Let $c_0, c_1, \ldots$ be a standard enumeration of C, and, as usual, let

$$C_s =_{def} \{c_0, c_1, \ldots, c_s\}$$

for each s.

We will construct incomparable, c.e. sets A and B such that $A \leq_T C$ and $B \leq_T C$, which suffice to prove the theorem (see Exercise 6 of Chap. 5).

The requirements and their relative priorities are exactly the same as in the Friedberg-Muchnik proof:

$$R_{2e} : \chi_A \neq \Phi_e^B$$
$$R_{2e+1} : \chi_B \neq \Phi_e^A$$

and

$$R_0 \prec R_1 \prec \cdots .$$

Again, at each stage s, we give attention to the strongest priority requirement R_j that needs it. In the Friedberg-Muchnik algorithm, giving attention entailed:

(1) initializing all requirements of weaker priority than that of R_j,
(2) putting the witness n_j for R_j into A (assuming that j is even; otherwise we put it into B).

In the current construction, we could try to separate these two tasks. That is, we could do (1) during stage s, but delay (2) until C permits it, that is, until we reach a stage $t \geq s$ such that $c_t \leq n_j$. Thus, by the Permitting Lemma, we would have $A \leq_T C$ and $B \leq_T C$.

Actually, it's more complicated than that. The trouble with this simple approach is that n_j might never be permitted. That is, perhaps $C_s \upharpoonright n_j = C \upharpoonright n_j$.

So, we will describe a more complicated approach, which is formalized as Algorithm 8.1. Before doing so, we define what is means for a witness to be "valid," and the three "types" of witnesses.

8.3 Valid Witnesses

From here until the end of this book, we write $A(k)$ to denote the more cumbersome $\chi_A(k)$. Analogously, we write simply A to denote the function χ_A. Using this notation, the requirements are

$$R_{2e} : A \neq \Phi_e^B$$
$$R_{2e+1} : B \neq \Phi_e^A .$$

Thus, A denotes either a set, or its characteristic function; the context will easily disambiguate it.

A witness n for R_{2e} is *valid*[1] if

$$A(n) \neq \Phi_e^B(n). \tag{8.3}$$

---

[1] When introducing the Friedberg-Muchnik algorithm in Chap. 7, we used the phrase "potential witness" to denote a number that may turn out to be a valid witness in this sense. We were tempted to distinguish between a "potential witness" and a "witness" (i.e., a potential witness that satisfies the full requirement, such as in (8.3)) throughout the remainder of this book. However, the phrase "potential witness" is too cumbersome, so instead we will distinguish between "witness" and "valid witness".

Some witness n might satisfy

$$A(n)[s] \neq \Phi_e^B(n)[s]$$

for some stage s, but that does not imply (8.3).

Likewise, a witness for R_{2e+1} is *valid* if

$$B(n) \neq \Phi_e^A(n).$$

8.4   Types of Witnesses

We define three types of witnesses n for requirement R_j at stage s (that is, at the start of stage s). They are specified in Fig. 8.1, where we assume that $j = 2e$ (for odd j we would interchange the roles of A and B).

Note that a Type 1 witness is valid if it *holds up*, that is, $n \notin A$ and $\Phi_e^B(n) \neq 0$. Likewise, a Type 3 witness is valid if it *holds up*, that is, $\Phi_e^B(n) = 0$ ($n \in A$ follows from $n \in A_s$). On the other hand, a Type 2 witness is useless if it *holds up*, that is, $n \notin A$ and $\Phi_e^B(n) = 0$. Our hope for a Type 2 witness is that it will be upgraded to Type 3 (that is, we put it into A, which requires the "permission" of C).

In the Friedberg-Muchnik algorithm, procedure *Initialize* threw away the current witness n_j for R_j, which could be of Type 1, 2, or 3, and replaced it by a fresh witness n_j. Recall that "fresh" means greater than every witness and every usage number that has been seen so far. If n_j was a Type 2 witness, then we said that R_j "needed attention." During each stage, if there was a requirement needing attention, then we gave attention to such a requirement of strongest priority. Giving attention to a requirement meant putting its witness into A (thereby upgrading it to Type 3), and initializing all weaker priority requirements. Thus, at all times, no requirement ever had more than one witness.

In Algorithm 8.1, requirement R_j maintains three witnesses, which we call $n_j(1)$, $n_j(2)$, and $n_j(3)$. At the end of each stage, for each $r \in \{1, 2, 3\}$, variable $n_j(r)$ is either undefined (in which case $n_j(r) = -1$), or it is a Type r witness.

If $n_j(1)$ never gets upgraded to Type 2 (that is, it holds up as Type 1), then R_j is met. Witness $n_j(2)$ is awaiting permission to enter A (in other words, it is awaiting an upgrade to Type 3).

Fig. 8.1 Three types of witnesses

Type	status at stage s
1	$n \notin A_s$ and $\Phi_e^B(n)[s] \neq 0$
2	$n \notin A_s$ and $\Phi_e^B(n)[s] = 0$
3	$n \in A_s$ and $\Phi_e^B(n)[s] = 0$

8.5 The Algorithm

ALGORITHM 8.1

```
 1   A ← ∅.
 2   B ← ∅.
 3   n₀(1) ← 0.        // In other words, n₀(1) ← a fresh number.
 4   n₀(2) ← −1.
 5   n₀(3) ← −1.
 6   for s ← 1 to ∞
 7       nₛ(1) ← a fresh number.
 8       nₛ(2) ← −1.
 9       nₛ(3) ← −1.
10       for j ← 0 to s
11           if c_{s+1} ≤ n_j(2)    // n_j(2) has permission to enter A.
                                     // Requirement R_j acts.
12               Put n_j(2) into A.
13               n_j(3) ← n_j(2).
14               n_j(2) ← −1.
15               for k ← j + 1 to s
16                   n_k(1) ← a fresh number.
17                   n_k(2) ← −1.
18                   n_k(3) ← −1.
19           if n_j(1) is a Type 2 witness and n_j(3) = −1
20               n_j(2) ← n_j(1).
21               n_j(1) ← a fresh number.
```

Notes on the algorithm:

1. The body of the <u>for</u> loop of line 10 is written as though j were even. If j is odd, then replace A by B.
2. The condition $n_j(3) = -1$ in line 19 prevents the upgrading of a Type 1 witness to Type 2 when R_j currently has a Type 3 witness.
3. Suppose that $n_j(1)[s]$ is upgraded to Type 2 (lines 19–21) during stage s, and suppose also that R_j had a Type 2 witness at the start of stage s (in other words, $n_j(2)[s] \geq 0$). Then we could keep both $n_j(1)[s]$ and $n_j(2)[s]$, and so would have two Type 2 witnesses awaiting an upgrade to Type 3. However, $n_j(2)[s] < n_j(1)[s]$; hence, if for some $t > s$,

$$c_{t+1} \leq n_j(2)[s]$$

(and so $n_j(2)[s]$ would receive permission to enter A during stage t), then

$$c_{t+1} \leq n_j(1)[s]$$

(and so $n_j(1)[s]$ would receive permission to enter A during stage t, too). Therefore we might as well keep only $n_j(1)[s]$ as a Type 2 witness, and throw away $n_j(2)[s]$. Hence we assign $n_j(1)$ to $n_j(2)$ (line 20) during stage s.

4. Two apparent difficulties with Algorithm 8.1 are:

   (a) Suppose that $n_j(1)[s]$ gets upgraded to Type 2 during stage s, and for some $t > s$,

   $$c_{t+1} \leq n_j(2)[s] \quad \text{and} \quad \Phi_e^B\big(n_j(1)[s]\big) \neq 0 \quad \text{and} \quad \Phi_e^B\big(n_j(2)[s]\big) = 0,$$

   where $e = \lfloor j/2 \rfloor$. In other words, $n_j(1)[s]$ does not hold up as a Type 3 witness, whereas $n_j(2)[s]$ would have, if we had kept it and upgraded it to Type 3 during stage t. Then wouldn't we regret our decision to throw away $n_j(2)[s]$ in line 20 during stage s? No, everything will work out fine, because

   $$\Phi_e^B\big(n_j(1)[s]\big)[s] = 0 \neq \Phi_e^B\big(n_j(1)[s]\big)$$

   implies that for some $i < j$, requirement R_i gets injured during or after stage s, which cannot happen for sufficiently large s, as follows from Lemma 8.3.

   (b) Suppose that for a particular j, requirement R_j never acts, but lines 20 and 21 are executed infinitely often. In other words, the variable $n_j(2)$ is assigned an infinite sequence of increasing values, none of which ever obtains permission to enter A. If so, then R_j would not be met. We will see in the proof of Lemma 8.4 that this cannot happen (because otherwise C would be computable).

8.6   Verification

Because A and B are c.e., Theorem 8 follows from Lemmas 8.2 and 8.4.

Lemma 8.2 $A \leq_T C$ and $B \leq_T C$.

Proof   To see that $A \leq_T C$, fix s and n, and assume that

$$n \in A_{s+1} - A_s.$$

In other words, during stage s of Algorithm 8.1, n is put into A by line 12. Hence, by line 11 of Algorithm 8.1, $c_{s+1} \leq n$. Thus,

$$c_{s+1} \in C_{s+1} \upharpoonright n - C_s \upharpoonright n$$
$$\subseteq C \upharpoonright n - C_s \upharpoonright n$$

and so

$$C_s \upharpoonright n \neq C \upharpoonright n.$$

Therefore (8.1) is true, and so, by the Permitting Lemma (Lemma 8.1), we have $A \leq_T C$. Analogous reasoning proves $B \leq_T C$.

Lemma 8.3 *Suppose that j and w are such that after stage w, variable $n_j(1)$ is never given a fresh number by line 16 of Algorithm 8.1. Then requirement R_j acts at most once after stage w.*

We leave the proof of Lemma 8.3 as Exercise 1.

Lemma 8.4 *For each j, requirement R_j acts only finitely often, and is met.*

Proof We proceed by strong induction on j.

For the basis, consider $j = 0$. Variable $n_0(1)$ is never given a fresh value by line 16 of Algorithm 8.1. Therefore, letting $w = 0$, Lemma 8.3 implies that R_j acts at most once after stage w.

For the inductive step, fix $j \geq 1$. Assume the claim for each $i < j$. Then there is a stage $w > j$ such that

$$(\forall i < j)[\, R_i \text{ never acts after stage } w\,].$$

Therefore, after stage w, variable $n_j(1)$ is never given a fresh number by line 16 of Algorithm 8.1. So, again, Lemma 8.3 implies that R_j acts at most once after stage w.

Everything we say from here to the end of the proof applies both to the basis and to the inductive step.

Because R_j acts at most once after stage w, R_j acts only finitely often. Hence there is an $x > w$ such that

$$(\forall i \leq j)[\, R_i \text{ never acts after stage } x\,].$$

It remains to show that R_j is met.

After stage x, every change to $n_j(2)$ is caused by line 20 of Algorithm 8.1, and hence is an increase (see Exercise 2). Therefore $n_j(2)$ never decreases after stage x. We claim that $n_j(2)$ is updated only finitely often; assume the contrary. Then

$$\max_s n_j(2)[s] = \infty. \tag{8.4}$$

Then the following algorithm, which knows j and x,[2] could determine whether a given p is in C:

---

[2] If you're a computer programmer, think of j and x as "hard-coded" constants (also known as "magic numbers"), whereas p is an input passed to Algorithm 8.2.

ALGORITHM 8.2

1 Run Algorithm 8.1 until reaching the first stage $s > \max\{x, j\}$ such that $n_j(2)[s] > p$.
2 **if** $p \in C_s$
3   output("p is in C")
4 **else** output("p is not in C").

Note that, by (8.4), the stage s will indeed be found by line 1 of Algorithm 8.2.

Assume for a contradiction that the output of Algorithm 8.2 were incorrect, that is, assume

$$p \in C - C_s.$$

Then there exists $t \geq s$ such that $p = c_{t+1}$. Thus,

$$
\begin{aligned}
c_{t+1} &= p \\
&< n_j(2)[s] \quad \text{(by line 1 of Algorithm 8.2)} \\
&\leq n_j(2)[t] \quad \text{(because } n_j(2) \text{ never decreases after stage } x\text{).}
\end{aligned}
$$

Therefore, because $t > j$, requirement R_j acts during stage t, contradicting $t > x$.

Thus, the output of Algorithm 8.2 is correct, and so C is computable, contrary to the assumption in the statement of the theorem. Therefore, $n_j(2)$ is updated only finitely often.

Let $y > x$ be a stage during or after which $n_j(2)$ is never updated. Then $n_j(1)$ never changes during or after stage y (because, whenever Algorithm 8.1 assigns to $n_j(1)$, it also assigns to $n_j(2)$).

Assume that $j = 2e$; if j is odd then the proof is analogous.

Case 1. $n_j(3)[y] \geq 0$.
Let $n = n_j(3)[y]$. No requirement R_i such that $i < j$ acts after stage w; hence, the computation $\Phi_e^B(n)[y]$ is permanent. Therefore n holds up as a Type 3 witness. Thus,

$$A(n) = 1 \neq 0 = \Phi_e^B(n)[y] = \Phi_e^B(n),$$

and so R_j is met.

Case 2. $n_j(3)[y] = -1$.
Let $n = n_j(1)[y]$.
Note that

$$\big(\forall y' \geq y\big)\big[n_j(3)[y'] = -1\big], \tag{8.5}$$

because R_j does not act during or after stage y.

We claim that

$$0 \neq \Phi_e^B(n). \tag{8.6}$$

To see this, assume the contrary. Then, by the Permanence Lemma,

$$\Phi_e^B(n)[z] = 0$$

for sufficiently large z. Hence, by (8.5), n would eventually become Type 2, causing $n_j(2)$ and $n_j(1)$ to be changed by lines 20 and 21 of Algorithm 8.2 during some stage $z' \geq y$, contradicting the choice of y. Thus, (8.6) is true. Furthermore, by (8.5), n never gets upgraded to Type 3; hence

$$A(n) = 0. \tag{8.7}$$

Combining (8.7) and (8.6), we have

$$A(n) = 0 \neq \Phi_e^B(n),$$

and so R_j is met.

QED Lemma 8.4

QED Theorem 8

8.7 What's New in This Chapter?

1. A simple form of permitting, which is a technique for constructing a c.e. set or sets below a given c.e.n. set C, while satisfying certain other properties as well.
2. Maintaining three possible witnesses rather than just one witness for a requirement. In Chap. 14, we generalize this by maintaining an arbitrarily long "witness list" for each requirement, which facilitates the more complicated form of permitting employed there.

   It is possible to implement Algorithm 8.1 with only two (rather than three) possible witnesses per requirement, but the pseudo-code would be more difficult to understand. Can it be implemented with only one possible witness per requirement? I don't know.
3. An algorithm (in particular, Algorithm 8.2) that is introduced as part of a proof by contradiction, and that knows as a "magic number" a certain stage (which is x, in Algorithm 8.2) after which certain finitely occurring events have ceased to occur. We will see this type of algorithm again, as Algorithms 9.2, 11.2, 12.2, and 13.2. Algorithm 14.2 is provided with both a magic number and an oracle. Chapter 14 combines most of the ideas of this book, and introduces a few new ones, too.

8.8 Afternotes

There are many varieties of permitting. Lemmas 8.1 and 8.5 (see Exercise 4) both fall into the category of "Yates permitting." There is also "Martin permitting" (which is explained in Chap. 11 of [So87]).

We use another variety of permitting in Chap. 14.

8.9 Exercises

1. Prove Lemma 8.3.
2. Prove that whenever line 20 of Algorithm 8.1 is executed, the variable $n_j(2)$
   increases.
3. Let C be c.e.n. Show that there is an infinite sequence

$$A_0 <_T A_1 <_T \cdots$$

   of c.e. sets such that

$$(\forall i)[\, A_i <_T C \,].$$

4. Prove the following generalization of Lemma 8.1:

Lemma 8.5 (Strong Permitting Lemma) *Let $A_0 \subseteq A_1 \subseteq \cdots$ and $C_0 \subseteq C_1 \subseteq \cdots$ be computable enumerations of A and C, respectively. Let $f : \omega \to \omega$ be a computable total function. If*

$$(\forall s, n)\big[\, n \in A_{s+1} - A_s \implies C_s \upharpoonright f(n) \neq C \upharpoonright f(n) \,\big]$$

then $A \leq_T C$.

Note that Lemma 8.1 is the special case of Lemma 8.5 in which f is the identity function.

3. In addition to the three types of witnesses defined in Fig. 8.1, we could define two more types, as in Fig. 8.2.

   (a) If a Type 4 witness holds up, would it be valid?
   (b) If a Type 5 witness holds up, would it be valid?

Fig. 8.2 Five types of witnesses

Type	status at stage s
1	$n \notin A_s$ and $\Phi_e^B(n)[s] \neq 0$
2	$n \notin A_s$ and $\Phi_e^B(n)[s] = 0$
3	$n \in A_s$ and $\Phi_e^B(n)[s] = 0$
4	$n \notin A_s$ and $\Phi_e^B(n)[s] = 1$
5	$\Phi_e^B(n)[s] > 1$

Chapter 9
Length of Agreement (Sacks Splitting Theorem)

The length of agreement method is widely used in the construction of c.e. sets. Suppose that we are given a c.e.n. set A, and we wish to construct a c.e.n. set B having certain properties, including $A \not\leq_T B$. This situation is very different from that in the proof of the Friedberg-Muchnik Theorem, because there we were building both A and B. Here, we build B but we have no control over A. We cannot put a number into A (to serve as a witness), nor can we keep anything out of A.

9.1 The Idea

Suppose that among our requirements are

$$R_e : A \neq \Phi_e^B$$

for each e. Assume that we can preserve whatever halting computations that we want. That is, if $\Phi_e^B(n)[s] \downarrow$ for some n and s, then we can restrain all numbers less than or equal to $\varphi_e^B(n)[s]$ from entering B during stage s or beyond. So, for example, if $\Phi_e^B(n)[s] = 42$ then we could lock in that computation and hence

$$A(n) \neq 42 = \Phi_e^B(n),$$

and so n would be a valid witness for requirement R_e, as defined in Sect. 8.3. Likewise, if $\Phi_e^B(n)[s] = 0$ for some s and some $n \in A_s$, then we could lock in that computation and hence

$$A(n) = 1 \neq 0 = \Phi_e^B(n),$$

© The Author(s), under exclusive license to Springer Nature Switzerland AG 2023
K. J. Supowit, *Algorithms for Constructing Computably Enumerable Sets*,
Computer Science Foundations and Applied Logic,
https://doi.org/10.1007/978-3-031-26904-2_9

Fig. 9.1 The length of agreement is 6

					k				
	0	1	2	3	4	5	6	7	8
$A_s(k)$	0	0	1	0	1	1	1	0	0
$\Phi_e^B(k)[s]$	0	0	1	0	1	1	$\uparrow$	0	17

Fig. 9.2 The length of agreement is 4

					k				
	0	1	2	3	4	5	6	7	8
$A_s(k)$	0	0	1	0	0	1	1	0	0
$\Phi_e^B(k)[s]$	0	0	1	0	1	1	$\uparrow$	0	17

and so again n would be a valid witness for R_e. However, it might be that neither of these fortuitous events ever happens. For example, it might be that for some stage s,

$$n \in A_s \quad \text{and} \quad \Phi_e^B(n)[s] \uparrow$$

but

$$\Phi_e^B(n) = 1. \tag{9.1}$$

Thus, n would be useless as a witness for R_e. Hence, we should replace n with another witness for R_e, but how could we know that (9.1) is true?

Therefore we employ the following, more powerful technique. Let

$$\ell_e(s) =_{def} \max \left\{ n : \left(\forall k < n \right) \left[A_s(k) = \Phi_e^B(k)[s] \right] \right\},$$

which is called the *length of agreement* between A and Φ_e^B at stage s. In Fig. 9.1, $\ell_e(s) = 6$.

In Fig. 9.2, $\ell_e(s) = 4$.

Note that $\Phi_e^B(k)[s] \downarrow$ for each $k < \ell_e(s)$. We try to preserve the computation $\Phi_e^B(k)[s]$ for each $k < \ell_e(s)$, so as to preserve the initial run of agreements. Furthermore, it is important that we likewise try to preserve the computation $\Phi_e^B(\ell(s))[s]$ if $\Phi_e^B(\ell(s))[s] \downarrow$ (as in Fig. 9.2), so as to preserve the first disagreement along with the initial run of agreements.

Recall that in the Friedberg-Muchnik algorithm, when a requirement R_j received attention, we put its witness n_j into A, and then tried to preserve the disagreement

$$A(n_j) = 1 \neq 0 = \Phi_e^B(n_j)[s]$$

by keeping small numbers out of B. In the length-of-agreement method, we try to preserve not only the first disagreement (if $\Phi_e^B(\ell_e)[s] \downarrow$), but also the initial run of agreements. This is why some authors refer to the length-of-agreement method as the "Sacks preservation strategy."

We will prove that the length of agreement cannot grow forever; that is, we will prove

$$\max_s \ell_s(s) < \infty. \tag{9.2}$$

In particular, we will argue that if

$$\max_s \ell_s(s) = \infty,$$

then, because of the preservation of agreements, A would be computable (contrary to our assumption that A is c.e.n.). Using (9.2), we will argue that R_e has a valid witness. This is just an overview; the argument is fleshed out in the following application.

9.2 The Theorem

The following result implies both the splitting theorem of Chap. 4 and the Friedberg-Muchnik Theorem of Chap. 7:

Theorem 9 (Sacks Splitting Theorem) *Let A be c.e.n. Then A can be partitioned into incomparable, c.e. sets B_0 and B_1.*

Proof We describe an algorithm that partitions A into incomparable, c.e. sets B_0 and B_1.

Recall Exercise 4(b) of Chap. 7. That problem and our current one are ostensibly similar. The only difference is that here we cannot ignore any elements of A; each member of A must be put either into B_0 or into B_1. This difference is huge; we cannot simply use the Friedberg-Muchnik approach of taking fresh witnesses, because in doing so we might skip over some elements of A.

So, instead of building B_0 and B_1 to meet the Friedberg-Muchnik requirements $B_0 \neq \Phi_e^{B_1}$ and $B_1 \neq \Phi_e^{B_0}$, so as directly to obtain

$$B_0 \not\leq_T B_1 \quad \text{and} \quad B_1 \not\leq_T B_0, \tag{9.3}$$

we will meet the requirement

$$R_{e,i} : A \neq \Phi_e^{B_i}$$

for each $e \in \omega$ and $i \in \{0, 1\}$. Collectively, the $R_{e,i}$ requirements ensure

$$A \not\leq_T B_0 \text{ and } A \not\leq_T B_1,$$

which imply (9.3), by the following lemma:

Lemma 9.1 *Assume that $A = B_0 \cup B_1$. Then*

$$A \not\leq_T B_0 \implies B_1 \not\leq_T B_0$$

and

$$A \not\leq_T B_1 \implies B_0 \not\leq_T B_1.$$

Proof To show $A \not\leq_T B_0 \implies B_1 \not\leq_T B_0$, we prove its contrapositive:

$$B_1 \leq_T B_0 \implies A \leq_T B_0.$$

Assume that $B_1 \leq_T B_0$ and that we have an oracle for B_0. Then, given p, we could determine whether

 (i) $p \in B_0$ by using the oracle for B_0, and
 (ii) $p \in B_1$ by using the Turing reduction from B_1 to B_0, together with the oracle for B_0.

These two tests suffice to determine whether $p \in A$, because $A = B_0 \cup B_1$.

The proof that $A \not\leq_T B_1 \implies B_0 \not\leq_T B_1$ is completely analogous.   □

On the face of it, the $R_{e,i}$ requirements might seem harder to meet than the Friedberg-Muchnik requirements, because we have no control over A. However, the length-of-agreement technique comes to our rescue.

The requirements are prioritized as

$$R_{0,0} \prec R_{0,1} \prec R_{1,0} \prec R_{1,1} \prec R_{2,0} \prec R_{2,1} \prec \cdots.$$

Let $a_0, a_1, \ldots$ be a standard enumeration of A. As in our other partitioning algorithm (Algorithm 4.1), during each stage s, we put a_s into B_0 or B_1 (but not into both), thereby ensuring that B_0 and B_1 partition A. Each stage is computable in a finite amount of time, and so both B_0 and B_1 are c.e.

9.3   Definitions

For each e and s, and each $i \in \{0, 1\}$, let

$$\ell_{e,i}(s) =_{def} \max \left\{ n \leq s : (\forall k < n)\left[A_s(k) = \Phi_e^{B_i}(k)[s]\right] \right\}.$$

That is, $\ell_{e,i}(s)$ equals either the length of agreement between A and $\Phi_e^{B_i}$ at the start of stage s, or it equals s, whichever is less. The reason for placing an upper bound (of s) on $\ell_{e,i}(s)$ is to ensure that it is finite, and that each stage of the algorithm can be performed in a finite amount of time.

Let

$$restraint_{e,i}(s) =_{def} \max \left\{ \varphi_e^{B_i}(k)[r] : k \leq \ell_{e,i}(r) \text{ and } r \leq s \right\}.$$

The algorithm attempts to keep numbers less than or equal to $restraint_{e,i}(s)$ from entering B_i during or after stage s and thereby possibly diminishing the length of agreement $\ell_{e,i}$. Note that $restraint_{e,i}(s)$ is monotonically non-decreasing in s.

Also note that we write "$k \leq \ell_{e,i}(r)$" rather than just "$k < \ell_{e,i}(r)$" in the definition of $restraint_{e,i}(s)$, so as to try to preserve the first disagreement along with the initial run of agreements, as discussed in Sect. 9.1.

9.4 The Algorithm

Algorithm 9.1 differs from Algorithm 4.1 only in lines 7 and 8.

ALGORITHM 9.1

```
 1  B₀ ← ∅.
 2  B₁ ← ∅.
 3  for s ← 0 to ∞
        // Assign aₛ either to B₀ or to B₁.
 4      assigned ← FALSE.
 5      for e ← 0 to s
 6          for i ← 0 to 1
 7              if (not assigned) and (aₛ ≤ restrainte,i(s))
                    // Re,i acts.
 8                  Put aₛ into B₁₋ᵢ.   // This keeps aₛ out of Bᵢ,
                                        // and so prevents aₛ from diminishing ℓe,i.
 9                  assigned ← TRUE.
10      if not assigned
11          Put aₛ into B₀.   // Or put aₛ into B₁, it doesn't matter.
```

9.5 Verification

For each s, the value $restraint_{e,i}(s)$ can be computed in a finite amount of time, and so stage s can be performed in a finite amount of time. Therefore B_0 and B_1 are c.e.

Lemma 9.2 *For each* $e \in \omega$ *and each* $i \in \{0, 1\}$,

   I. $\max_s \ell_{e,i}(s) < \infty$,
   II. $\max_s restraint_{e,i}(s) < \infty$,
   III. $R_{e,i}$ *acts only finitely often.*

Proof We use strong induction on the priority of $R_{e,i}$. The inductive step is as follows.[1] Fix e and i, and assume the claim for each e' and i' such that

$$R_{e',i'} \prec R_{e,i}.$$

Then (by Part III) there is a stage x after which no requirement of stronger priority than $R_{e,i}$ ever acts (informally, after stage x, requirement $R_{e,i}$ is treated like $R_{0,0}$). Therefore, for each $t > x$, if

$$a_t \leq restraint_{e,i}(t)$$

then $R_{e,i}$ acts during stage t, putting a_t into B_{1-i}. Hence

$$(\forall t > x)\big[\, a_t \in B_i \implies a_t > restraint_{e,i}(t) \,\big]. \tag{9.4}$$

Proof of Part I. Assume for a contradiction that Part I were false; that is, assume

$$\max_s \ell_{e,i}(s) = \infty. \tag{9.5}$$

Then the following algorithm, which knows x, could determine whether a given p is in A:

ALGORITHM 9.2

1 Run Algorithm 9.1 until reaching the first
        stage $s > x$ such that $\ell_{e,i}(s) > p$.   // Such an s exists by (9.5).
2 **if** $\Phi_e^{B_i}(p)[s] = 1$
3   output("p is in A")
4 **else** output("p is not in A").

If $t \geq s$ and $a_t \in B_i$ then

$$a_t > restraint_{e,i}(t) \qquad \text{(by (9.4), because } t \geq s > x)$$
$$\geq restraint_{e,i}(s) \qquad \text{(by the monotonicity of } restraint_{e,i})$$
$$\geq \max \big\{\varphi_e^{B_i}(k)[s] : k \leq \ell_{e,i}(s)\big\} \quad \text{(by the definition of } restraint_{e,i}(s))$$
$$\geq \varphi_e^{B_i}(p)[s] \qquad \text{(because } p < \ell_{e,i}(s)).$$

Therefore

$$B_i[s] \!\upharpoonright\! \varphi_e^{B_i}(p)[s] = B_i \!\upharpoonright\! \varphi_e^{B_i}(p)[s],$$

---

[1] The argument for the basis, in which $e = i = 0$, is a special case of the inductive step. This was true in our priority arguments in Chaps. 7 and 8, where nevertheless we worked out the details of the bases. It is true also in the priority arguments in Chaps. 11 and 12, where we won't bother to even mention the bases.

and so the computation $\Phi_e^{B_i}(p)[s]$ is permanent, as defined in Chap. 6. Hence

$$\Phi_e^{B_i}(p) = \Phi_e^{B_i}(p)[s]. \tag{9.6}$$

If

$$\Phi_e^{B_i}(p)[s] = 1$$

then $p \in A_s$ (because $\ell_{e,i}(s) > p$), and so $p \in A$. On the other hand, if

$$\Phi_e^{B_i}(p)[s] = 0$$

then (by (9.6))

$$\Phi_e^{B_i}(p) = 0$$

and so $p \notin A$, because otherwise

$$A(p) = 1 \neq 0 = \Phi_e^{B_i}(p), \tag{9.7}$$

which would contradict (9.5) (see Exercise 3). To summarize,

$$p \in A \iff \Phi_e^{B_i}(p)[s] = 1,$$

and so the output of Algorithm 9.2 is correct. Thus A is computable, contrary to the assumption in the statement of the theorem.[2] Thus, Part I is proved.

Proof of Part II. Let $y > x$ be such that

$$\ell_{e,i}(y) = \max_{s>x}\{\ell_{e,i}(s)\}$$

(such a y exists by Part I). Let $\ell = \ell_{e,i}(y)$.
    If $z \geq y$ and $a_z \in B_i$ then

$$\begin{aligned} a_z &> restraint_{e,i}(z) & \text{(by (9.4), because } z \geq y > x) \\ &\geq restraint_{e,i}(y) & \text{(by the monotonicity of } restraint_{e,i}). \end{aligned}$$

In other words,

$$B_i[y] \upharpoonright restraint_{e,i}(y) = B_i \upharpoonright restraint_{e,i}(y).$$

Therefore

$$(\forall k < \ell)\big[\text{the computation } \Phi_e^{B_i}(\ell)[y] \text{ is permanent}\big]. \tag{9.8}$$

---

[2] This argument is reminiscent of (but more complicated than) the main argument in the proof of Lemma 8.4.

and so
$$\left(\forall z \geq y\right)\left[\,\ell_{e,i}(z) = \ell\,\right].$$

Hence if $z \geq y$ and $\Phi_e^{B_i}(\ell)[z] \downarrow$ then

$$restraint_{e,i}(z) \geq \varphi_e^{B_i}(\ell)[z],$$

and so the computation $\Phi_e^{B_i}(\ell)[z]$ is permanent.[3] Thus,

$$\left(\forall z \geq y\right)\left[\,\Phi_e^{B_i}(\ell)[z] \downarrow \implies \varphi_e^{B_i}(\ell) \text{ never changes during or after stage } z\,\right]. \quad (9.9)$$

Case 1. $\Phi_e^{B_i}(\ell)[y] \downarrow$.
   Then $\varphi_e^{B_i}(\ell)$ never changes during or after stage y, by (9.9).
Case 2. $\Phi_e^{B_i}(\ell)[y] \uparrow$ and $\Phi_e^{B_i}(\ell)[z] \downarrow$ for some $z > y$.
   Then $\varphi_e^{B_i}(\ell)$ changes exactly once during or after stage y, by (9.9).
Case 3. $\left(\forall z \geq y\right)\left[\,\Phi_e^{B_i}(\ell)[z] \uparrow\,\right]$.
   Then $\left(\forall z \geq y\right)\left[\,\varphi_e^{B_i}(\ell)[z] = -1\,\right]$.

In each of the three cases, $\varphi_e^{B_i}(\ell)$ changes at most once during or after stage y, and so, by (9.8),

$$\max\left\{\varphi_e^{B_i}(k) : k \leq \ell\right\}$$

changes at most once during or after stage y. Hence $restraint_{e,i}$ changes only finitely often, and so

$$\max_s restraint_{e,i}(s) < \infty.$$

Proof of Part III. By Part II, there exists r such that

$$r = \max_s restraint_{e,i}(s).$$

Let z be such that
$$A_z \upharpoonright r = A \upharpoonright r.$$

Then
$$\left(\forall s > z\right)\left[a_s > r \geq restraint_{e,i}(s)\right].$$

Therefore, for each $s > z$, the condition in line 7 of Algorithm 9.1 evaluates to false. Hence $R_{e,i}$ never acts after stage z, and so it acts only finitely often.

QED Lemma 9.2

---

[3] This is why we preserve the first disagreement along with the initial run of agreements.

Lemma 9.3 *For each $e \in \omega$ and each $i \in \{0, 1\}$, $R_{e,i}$ is met.*

Proof Fix e and i. Assume for a contradiction that $R_{e,i}$ is not met; that is,

$$A = \Phi_e^{B_i}. \tag{9.10}$$

By Part I of Lemma 9.2, there exists

$$m = \max_s \{\, \ell_{e,i}(s) \,\}. \tag{9.11}$$

By (9.10) and the Permanence Lemma, there is an x such that

$$\big(\forall k \le m\big)\big[\text{the computation } \Phi_e^{B_i}(k)[x] \text{ is permanent}\big].$$

Let $y > x$ be such that

$$\big(\forall k \le m\big)\big[\, A_y(k) = A(k)\,\big].$$

By (9.10), we have

$$\ell_{e,i}(y) > m,$$

contradicting (9.11).

QED Lemma 9.3

The theorem follows from Lemmas 9.1 and 9.3.

QED Theorem 9

9.6 Why Preserve Agreements?

The preservation of agreements is counter-intuitive. Is it necessary?

Suppose that during a stage $s > x$ (where x is as defined in Lemma 9.2), we find some k such that

$$\Phi_e^{B_i}(k)[s] > 1. \tag{9.12}$$

or

$$\Phi_e^{B_i}(k)[s] = 0 \quad \text{and} \quad k \in A_s. \tag{9.13}$$

Then we need only preserve the computation $\Phi_e^{B_i}(k)[s]$, to make k a valid witness for $R_{e,i}$ (because if (9.12) is true then it doesn't matter whether $k \in A$, and if (9.13) is true then $k \in A$).

Instead of the function $restraint_{e,i}(s)$ defined in Sect. 9.3, suppose that we define

$$restr_{e,i}(s) =_{def} \begin{cases} \varphi_e^{B_i}(k)[s], & \text{if } k \text{ is the least number satisfying either (9.12) or (9.13) at stage } s \\ -1, & \text{if no such } k \text{ exists at stage } s \end{cases}$$

and replace the clause $a_s \leq restraint_{e,i}(s)$ in line 7 of Algorithm 9.1 by $a_s \leq restr_{e,i}(s)$. Unlike $restraint_{e,i}(s)$, the simpler function $restr_{e,i}(s)$ tries to preserve only a single disagreement; it does not try to preserve agreements. The following is an invalid argument that this modified algorithm still works (that is, it constructs sets B_0 and B_1 such that each $R_{e,i}$ is met).

Fix e and i. First, we "claim" that

$$\liminf_s \ell_{e,i}(s) < \infty; \tag{9.14}$$

assume for a contradiction that

$$\liminf_s \ell_{e,i}(s) = \infty. \tag{9.15}$$

Note that for a function $f : \omega \to \omega$ (such as $\ell_{e,i}(s)$),

$$\liminf_s f(s) = \min\{k : f(s) = k \text{ for infinitely many } s\}, \tag{9.16}$$

where the minimum is regarded as ∞ if the set is empty. This is a convenient way to think about the liminf, and is used extensively in Chap. 14.

By (9.15) and (9.16), there exists y such that

$$(\forall s \geq y)[\ell_{e,i}(s) > p].$$

Then the following algorithm, which knows y, could determine whether a given p is in A:

> ALGORITHM 9.2′
>
> Run Algorithm 9.1 until reaching stage y.
> **if** $\Phi_e^{B_i}(p)[y] = 1$
>   output("p is in A")
> **else** output("p is not in A").

Algorithm 9.2′, given p, correctly determines whether $p \in A$. Therefore A is computable, contrary to the assumption in the statement of the theorem.

Now that we have "proven" (9.14), let

$$n = \liminf_s \ell_{e,i}(s).$$

Then

$$\Phi_e^{B_i}(n)\uparrow,$$

by the contrapositive of the Permanence Lemma, and so n is a valid witness for $R_{e,i}$.

QUESTION: What is the flaw in this argument? Perhaps think about it, before you read the next paragraph.

ANSWER: In the proof of Part I of Lemma 9.2, x was defined independently of the input p; this justified our saying that Algorithm 9.2 "knows x." That is, x was part of the description of the algorithm. In this invalid argument, y depends on p (it should be called y_p), and so Algorithm 9.2' cannot know y_p. Instead, given p, Algorithm 9.2' would need to compute y_p, which might be impossible to do.

9.7 What's New in This Chapter?

1. The length-of-agreement method, also known as the Sacks preservation strategy. We'll employ it, in various forms, in most of the subsequent chapters.
2. In Algorithms 7.1 and 8.1, for each requirement we chose a witness, and then monitored it. Sometimes we upgraded it (in Algorithm 8.1), sometimes we put it into an appropriate set (A or B), and sometimes we replaced it by a fresh witness. Algorithm 9.1, on the other hand, does not explicitly choose or even look at any witnesses; hence, it has no procedure *Initialize*. Rather, for each requirement we pay attention only to its *restraint* function. We rely on the length-of-agreement method to produce a witness without telling us about it.

9.8 Afternotes

Theorem 9 is actually a slightly weaker version of what is known in the literature as the "Sacks Splitting Theorem," first proved in [Sa63]. See [DS] for a survey of splitting theorems, including (but not limited to) various generalizations of Theorems 4 and 9.

We found in Exercise 2 of Chap. 7 an explicit form (as a function of j) for an upper bound on the number of times that requirement R_j is injured during the Friedberg-Muchnik algorithm. Here in the Sacks Splitting algorithm, the number of injuries to $R_{e,i}$ is indeed finite, but is there a nice form for its upper bound (as a function of e and i)? Is such a bound even computable?

Can the Sacks Splitting Theorem be proved without a length-of-agreement argument?

9.9   Exercises

1. Can a computable set be partitioned into incomparable c.e. sets B_0 and B_1?
2. Recall (9.4):

$$(\forall t > x)\big[\, a_t \in B_i \implies a_t > restraint_{e,i}(t)\,\big].$$

Is it true that

$$(\forall t > x)\big[\, a_t \in B_i \impliedby a_t > restraint_{e,i}(t)\,\big]?$$

3. Prove that (9.7) contradicts (9.5).
4. Say whether the following sentence is true or false:

$$\big(\forall e \in \omega\big)\big(\forall i \in \{0, 1\}\big)\big[\, \lim_s \ell_{e,i}(s) \text{ exists}\,\big].$$

Prove your answer.
5. Let A be c.e.n. Can A can be partitioned into infinitely many c.e. sets $B_0, B_1, \ldots$ that are pairwise incomparable?
6. The sets B_0 and B_1 produced by Algorithm 9.1 are both non-computable, because if one of them were computable then it would Turing reduce to the other. Therefore, Theorem 9 is a stronger version of Theorem 4. Is there a more direct argument that they are non-computable, that is, an argument that does not first show $B_0 \not\leq_T B_1$ and $B_0 \not\leq_T B_1$?
7. Consider the following:
Theorem $9'$. Let A be c.e., and let C be c.e.n. Then A can be partitioned into c.e. sets B_0 and B_1 such that

$$C \not\leq B_0 \quad \text{and} \quad C \not\leq B_1.$$

(a) Prove that Theorem $9'$ implies Theorem 9.
(b) Prove Theorem $9'$.

8. Let C and A be c.e.n. Prove that A can be partitioned into incomparable c.e. sets B_0 and B_1 such that $B_0 \leq_T C$ and $B_1 \leq_T C$.
9. Show that if A is c.e.n. then A can be partitioned into incomparable, c.e.n. sets B_0 and B_1 such that B_0 is sparse (this is a stronger version of Exercise 4 of Chap. 4).

Chapter 10
Introduction to Infinite Injury

Coping with infinite injury is a subtle and beautiful topic. You should read this short introductory chapter now, and then again after studying Chap. 11. Many of the terms used in this chapter need definitions, which will be provided in Chap. 11.

10.1 A Review of Finite Injury Priority Arguments

In finite injury arguments, we construct a set B (or perhaps multiple sets such as B_0 and B_1), subject to certain requirements. A typical such argument has positive requirements P_e that "act" by putting certain elements into B, and negative requirements N_e that try to protect certain OTM computations from injury by restraining sufficiently small elements from entering B.[1] Thus, the positive and negative requirements must coexist in an adversarial relationship. The priorities maintain law and order, allowing all requirements to ultimately be met.

In particular, if s is a stage at the start of which a negative requirement N_e is met, then N_e tries to keep numbers $b \leq restraint_e(s)$ from entering B, where $restraint_e(s)$ is so large that numbers above it cannot harm the work of N_e. Sometimes N_e fails in this endeavor—that is, it gets "injured"—by the action of a positive requirement of stronger priority, say, P_{e-50}. This could result in $restraint_e(s+1) >$

---

[1] In the proof of the Friedberg-Muchnik Theorem, we did not label our requirements as P for positive and N for negative. Rather, each requirement, say $R_{2e} : A \neq \Phi_e^B$, was both positive and negative. That is, R_{2e} sometimes needed to force a witness into A, and sometimes needed to keep small elements out of B. That's why we used the neutral label R.

In the proof of the Sacks Splitting Theorem, requirement $R_{e,i}$ certainly was negative. It could also be viewed as positive in the sense that it might put certain elements into B_{1-i} (but for the sole purpose of keeping them out of B_i).

© The Author(s), under exclusive license to Springer Nature Switzerland AG 2023
K. J. Supowit, *Algorithms for Constructing Computably Enumerable Sets*,
Computer Science Foundations and Applied Logic,
https://doi.org/10.1007/978-3-031-26904-2_10

restraint_e(s), an irksome situation, because it might prevent a weaker priority positive requirement (such as P_{e+17}) from acting. However, in the priority arguments
that we have seen so far, we argued (by induction) that each positive requirement
acts only finitely often, and so N_e gets injured only finitely often, and hence

$$\max_s restraint_e(s) < \infty.$$

10.2 Coping with Infinite Injury

What if positive requirement P_{e-50} needs to act infinitely often, causing infinitely
many injuries to N_e? Such positive requirements appear in the proof of the Weak
Thickness Lemma in Chap. 11, and in other proofs in subsequent chapters. This could
cause

$$\max_s restraint_e(s) = \infty,$$

which might prevent P_{e+17} from acting as much as it needs to. At first glance,
the difficulties associated with infinite injury might seem insurmountable. However,
methods for coping with them have been developed.

10.2.1 Guessing

The main technique for coping with infinite injury, which is the only one studied in
this book, involves a "priority tree," which is a rooted infinite tree in which the nodes
on level e have the responsibility for meeting requirements P_e and N_e.[2] Each node on
level e of the tree represents a sequence of guesses, which is why we sometimes refer
to a priority tree as a "tree of guesses." In the simpler tree arguments (such as those
in Chaps. 11 and 12), these are guesses about the given set A. In some of the more
complicated tree arguments (such as those in Chaps. 13 and 14), these are guesses
about the constructed set(s) or about some internal variables of the algorithm.

For each node σ of the tree, we define a function $restraint_\sigma(s)$. Think of it as a
more refined version of $restraint_e(s)$, where σ is on level e of the tree, because it
depends not only on e but also on the specific guesses that are represented by σ.

During each stage s, a path, called TP_s, is constructed from the root of the tree
down to some level s node. We think of TP_s both as a path, and as the set of $s + 1$
nodes along that path. Each node $\sigma \in TP_s$ is said to be "visited" during stage s. When

---

[2] This is true for the first generation of priority tree arguments, sometimes called "double-jump"
arguments. For the next generation, the daunting "triple-jump" arguments (which are not treated in
this book), the relationship between level e and the requirements P_e and N_e can be more complicated.

σ is visited, it may act (the definition of action depends on the particular algorithm; for the simpler ones, action consists only of putting an element into B). However, when σ is visited, it must respect the restraints imposed by all nodes of stronger priority. "Respect" means that σ may not put an element into B that violates one of those restraints. A node ρ has "stronger priority" than σ if it lies to the left of σ, or above σ, in the tree (we'll be more precise about that in Chap. 11).

The set

$$\{\text{the leftmost node on level } e \text{ that is visited infinitely often} : e \in \omega\}$$

forms an infinite path, called the "true path." If σ lies to the left of the true path then σ is visited only finitely often, and hence

$$\max_{s} restraint_{\sigma}(s) < \infty. \tag{10.1}$$

It also turns out (and this is crucial) that if σ lies on the true path then it, too, satisfies (10.1). Furthermore, if σ lies on the true path then each of the guesses represented by σ is correct. These two properties of the true path are often the key to proving that all the requirements are met.

10.2.2 Other Methods

There were other methods (which have largely been replaced by priority trees) developed for coping with infinite injury. Among them are:

1. The first[3] method developed for coping with infinite injury involved identifying a set T of "true stages." It would be great if

$$\max_{t \in T} restraint_{e}(t) < \infty,$$

but that may not be the case. A modified (known as "hatted") restraint function $\hat{r}_{e}(s)$ is defined so that

$$\max_{t \in T} \hat{r}_{e}(t) < \infty.$$

The positive requirements are able to get their work done during these true stages. There is a fair amount of detail in the definitions of T and of $\hat{r}_{e}(s)$.

---

[3] Arguably, it traces back to 1954 [De]. True stages are not only an algorithmic alternative to priority trees; in [CGS], for example, the concept of true stages is used in the proof of correctness of a complicated priority tree algorithm. Various generalizations of true stages are presented in [Mo].

2. Then came the "pinball" method, which according to [So87], was "slightly more cumbersome but more versatile" than the method of true stages.

Which technique (true stages, pinball, or priority tree) is conceptually simplest? Which is most aesthetically pleasing? Opinions vary on this.

Chapter 11
A Tree of Guesses (Weak Thickness Lemma)

This chapter contains our first infinite injury argument.

11.1 The "Lemma"

For all A and e, let

$$A^{[e]} =_{def} \{\langle e, j \rangle \in A : j \in \omega\}.$$

Think of $A^{[e]}$ as *row e* of A. For example, assume that

$$A \restriction 20 = \{3, 5, 7, 10, 12, 14, 17, 18\}.$$

then we can view A as the set of circled elements in the matrix shown in Fig. 11.1.

Thus, in Fig. 11.1, we see that $17 \in A^{[3]}$, for example. Compare Fig. 11.1 with the figure in Appendix A. Note that if A is c.e., then so is $A^{[e]}$ for each e.

We say that $A^{[e]}$ is *full* if it equals $\omega^{[e]}$; pictorially, this means that each element of row e of A is circled. Thus, in Fig. 11.1, $A^{[2]}$ might be full, whereas $A^{[0]}$, $A^{[1]}$, $A^{[3]}$, $A^{[4]}$, and $A^{[5]}$ definitely are not. Set A is *piecewise trivial* if

$$(\forall e)\left[A^{[e]} \text{ is either full or finite}\right].$$

If $B \subseteq A$ then B is a *thick* subset of A if for all e, $A^{[e]} - B^{[e]}$ is finite (informally, $B^{[e]}$ almost equals $A^{[e]}$).

© The Author(s), under exclusive license to Springer Nature Switzerland AG 2023   77
K. J. Supowit, *Algorithms for Constructing Computably Enumerable Sets*,
Computer Science Foundations and Applied Logic,
https://doi.org/10.1007/978-3-031-26904-2_11

$e\backslash j$	0	1	2	3	4	5	$\cdots$
0	0	2	⑤	9	⑭	20	$\cdots$
1	1	4	8	13	19	$\cdots$	
2	③	⑦	⑫	⑱	$\cdots$		
3	6	11	⑰	$\cdots$			
4	⑩	16	$\cdots$				
5	15	$\cdots$					
$\vdots$							

Fig. 11.1   Viewing a set A as a matrix

Theorem 11   (Weak Thickness Lemma[1])

Let A be c.e.n. and piecewise trivial. Then there is a thick, c.e. subset B of A such that $A \not\leq_T B$.

Before proving Theorem 11, let's consider two examples, where C is a c.e.n. set:

1. $A = \{\langle e, 0\rangle : e \in C\}$. Then A is piecewise trivial and c.e.n. The set $B = \emptyset$ is a thick, c.e. subset of A such that $A \not\leq_T B$ (because B is computable, whereas A is not).
2. $A = \{\langle e, j\rangle : e \in C \text{ and } j \in \omega\}$. Then A is piecewise trivial and c.e.n. So, by Theorem 11, there is a thick, c.e. subset B of A such that $A \not\leq_T B$. However, it's not so easy to identify one.

Proof   As usual, let

$$a_0, a_1, \ldots$$

denote a standard enumeration of A, and for each s, let $A_s =_{def} \{a_0, a_1, \ldots, a_s\}$.

We construct B to meet the requirements

$$P_e : A^{[e]} - B^{[e]} \text{ is finite}$$
$$N_e : A \neq \Phi_e^B$$

for each e.

We never put an element into B unless we have already seen it in A, so we obtain $B \subseteq A$ without the need for additional requirements. If $A^{[e]}$ is full then, in order to meet P_e, we must put all but finitely many elements of $A^{[e]}$ into $B^{[e]}$. On the other

---

[1] See the comment in the Afternotes section about why this result is called a "lemma" in the literature but a "theorem" in this chapter.

hand, if $A^{[e]}$ is finite then P_e will automatically be met and so in principle we could ignore it; however, we have no algorithm to determine, for a given e, whether $A^{[e]}$ is full or finite, and so in practice we can never ignore P_e.

11.2 The Tree

We would like to prioritize the requirements as

$$N_0 \prec P_0 \prec N_1 \prec P_1 \prec \cdots$$

and use a length-of-agreement strategy to meet N_e, as we did in the proof of the Sacks Splitting Theorem. In that proof we tried to protect a computation $\Phi_e^{B_i}(k)[s]$ by preventing all $b \le \varphi_e^{B_i}(k)[s]$ from entering B_i during or after stage s and thereby injuring $R_{e,i}$. Sometimes this effort would fail, because a small number could enter B_i by the action of a stronger priority requirement. It all worked out because the stronger priority requirements acted only finitely often, and hence (by induction), all requirements acted only finitely often. In the current proof, on the other hand, if $A^{[e]}$ is full then P_e acts infinitely often; thus, for some e' such that $e < e'$, the actions of P_e might infinitely often injure $N_{e'}$.

However, if we knew in advance which rows of A were full, then we could proceed with the length-of-agreement strategy for the N-requirements. How to proceed is a bit complicated, and we will see the details soon. Of course, we do know not in advance which rows are full, but we can benefit by guessing. We organize our guesses as a tree, as depicted in Fig. 11.2. It is an infinite binary tree, in which each node's left branch is labeled ∞ and right branch is labeled f. Thus, we think of nodes as finite-length strings over the alphabet $\{\infty, f\}$. We write:

1. $|\sigma|$ to denote the length of (i.e., the number of symbols in) string σ. Thus, node σ lies on level $|\sigma|$ of the tree. For example, the empty string λ is the root of the tree, and $|\lambda| = 0$.
2. $\tau \preceq \sigma$ if node τ is an ancestor of node σ. Each node is an ancestor of itself, that is, $\sigma \preceq \sigma$. Viewed as strings rather than as tree nodes, $\tau \preceq \sigma$ means that τ is a prefix (that is, an initial substring) of σ. Recall that in Chap. 5 we wrote $\tau \prec \sigma$ to mean that string τ is a proper prefix of string σ; thus, when viewing strings as tree nodes, $\tau \prec \sigma$ means that τ is a proper ancestor of σ, that is, $\tau \preceq \sigma$ but $\tau \neq \sigma$.
3. $\rho <_L \sigma$ if
$$(\exists \tau)[\, \tau^\frown \infty \preceq \rho \quad \text{and} \quad \tau^\frown f \preceq \sigma \,],$$

as is depicted in Fig. 11.3; here we say that "ρ is to the left of σ" or "σ is to the right of ρ." In each drawing of a tree in this book, a straight line segment indicates a single edge, whereas a wavy line indicates a path of zero or more edges.

<u>level</u>

0

1

2

3

4

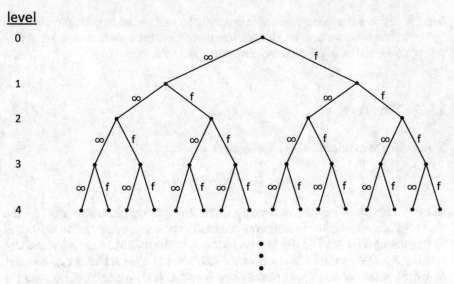

Fig. 11.2   The tree for Algorithm 11.1

Fig. 11.3 $\rho <_L \sigma$

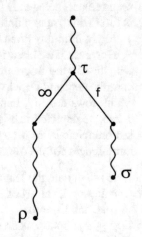

4. $\rho \lhd \sigma$ if

$$|\rho| = |\sigma| \quad \text{and} \quad \rho <_L \sigma,$$

that is, ρ and σ are on the same level of the tree, and ρ is to the left of σ.
5. $\rho \unlhd \sigma$ if

$$\rho \lhd \sigma \quad \text{or} \quad \rho = \sigma.$$

Intuitively, each node σ of the tree represents a sequence of $|\sigma|$ guesses. Consider the node $\sigma = f\infty\infty f$, for example, which is indicated in Fig. 11.4. This node σ represents the following $|\sigma| = 4$ guesses:

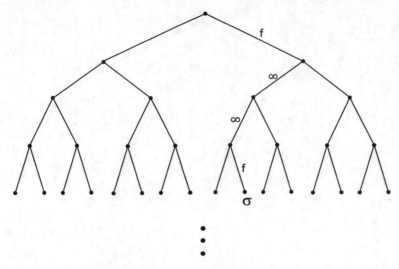

Fig. 11.4 The node $\sigma = f\infty\infty f$

1. $A^{[0]}$ is finite, because the 0th symbol of σ is f,
2. $A^{[1]}$ is full, because the 1st symbol of σ is ∞,
3. $A^{[2]}$ is full, because the 2nd symbol of σ is ∞,
4. $A^{[3]}$ is finite, because the 3rd symbol of σ is f.

During each stage s, Algorithm 11.1 computes TP_s, which is the set of nodes on the path from the root down to some node on level s of the tree. Thus, $|TP_s| = s + 1$. The following three statements are equivalent:

1. $\sigma \in TP_s$.
2. σ is *visited* during stage s.
3. s is a σ-*stage*.

Now, you might regard having three phrases to express the same concept as extravagant. We indulge in this luxury because certain sentences flow better using one of these phrases rather than the others, and because all three are ubiquitous in the literature.

A set P of nodes in this tree is an *infinite path* if $\lambda \in P$, and each node in P has exactly one child in P. Informally, an infinite path starts at the root and descends forever. Let

$$TP =_{def} \{\text{the leftmost node on level } e \text{ that is visited infinitely often} : e \in \omega\}.$$
$$(11.1)$$

Then TP is an infinite path. For each s, TP_s may or may not be a subset of TP_{s+1}, or of TP. The sequence

$$TP_0, TP_1, \ldots$$

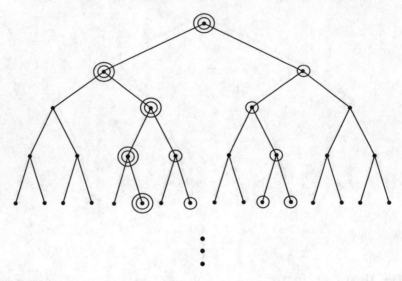

Fig. 11.5 The true path consists of the twice-circled nodes

may be thought of as a sequence of lightning bolts (each slightly longer than its predecessor), that zig and zag somewhat independently from each other.[2] The name *TP* abbreviates "true path," because all of the (infinitely many) guesses along *TP* are actually correct, as we will show as Lemma 11.1. In fact, *TP* is the only infinite path whose guesses are all correct, because distinct infinite paths $path_1$ and $path_2$ guess differently at the node α where they diverge (that is, $\alpha^\frown\infty \in path_1$ and $\alpha^\frown f \in path_2$, or vice versa).

In Fig. 11.5, each node circled either once or twice is visited infinitely often. Uncircled nodes are visited only finitely often. Nodes circled twice constitute the members of *TP*; each level has exactly one member of *TP*. If a node is circled once then each of its ancestors is circled either once or twice. If a node is circled twice then so is each of its ancestors.

If

$$(\exists \tau \in TP)[\, \sigma \lhd \tau \,]$$

then we write $\sigma <_L TP$ and say that "σ is to the left of *TP*." Likewise, if

$$(\exists \tau \in TP)[\, \tau \lhd \sigma \,]$$

then we write $TP <_L \sigma$ and say that "σ is to the right of *TP*." Thus, every node is either to the left of, in, or to the right of *TP*.

Analogously, if

$$(\exists \tau \in TP_s)[\, \sigma \lhd \tau \,]$$

---

[2] Some students have found this lightning bolt metaphor to be helpful. If you don't, then ignore it.

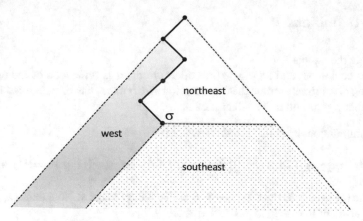

Fig. 11.6 The partition of the tree nodes induced by σ

then we write $\sigma <_L TP_s$ and say that "σ is to the left of TP_s." Likewise, if

$$(\exists \tau \in TP_s)[\tau \lhd \sigma]$$

then we write $TP_s <_L \sigma$ and say that "σ is to the right of TP_s."

Each node σ (whether or not $\sigma \in TP$) partitions all of the tree nodes into three sets:

1. $\{\rho : \rho <_L \sigma\}$. These nodes are *west* of σ.
2. $\{\tau : (\exists \alpha)[\sigma \unlhd \alpha \text{ and } \alpha \preceq \tau]\}$. These nodes are *southeast* of σ.
3. All the rest. These nodes are *northeast* of σ.

This terminology is illustrated in Fig. 11.6. Note that the southeast region includes all of its boundary nodes (including σ). The northeast region includes its western boundary, that is, the proper ancestors of σ.

Intuitively, each node σ on level e of the tree is trying to meet two requirements: P_e and N_e. To meet P_e, node σ must force certain elements of $A^{[e]}$ into B.

To meet N_e, node σ makes its own length-of-agreement argument. As part of this length-of-agreement argument, σ tries to prevent each $b \leq restraint_\sigma$ (which is defined in the next section) from entering B. The nodes southeast of σ must respect $restraint_\sigma$; that is, they may not put numbers $b \leq restraint_\sigma$ into B, and hence they cannot injure N_e. The nodes northeast of σ likewise do not injure N_e, but the proof in this case is trickier; this is where the guessing comes into play. The nodes west of σ need not respect $restraint_\sigma$, and so can indeed injure N_e; however, if $\sigma \in TP$ then such nodes act (and thereby possibly injure N_e) only finitely often.

For each e, all that we need is for at least one node on level e to succeed in meeting P_e and N_e. The node on level e of the true path will succeed (it is possible that some nodes to the right of the true path will succeed also, but we don't need them to).

Our tree may be thought of as a tree of guesses, but (as we said in Chap. 10), it is called a "priority tree" in the literature.

11.3 Definitions

Herein lie the details.

Let σ be a node, and let $e = |\sigma|$. In other words, σ is on level e of the tree. The following definitions assume that the function $Restraint_\rho$ has been defined for each ρ such that $\rho \neq \sigma$ and σ is southeast of ρ.

1. A computation $\Phi_e^B(k)[s]$ is σ-believable if

$$(\forall \tau^\frown \infty \preceq \sigma)(\forall a \in \omega^{[|\tau|]})\big[\, Restraint_\tau(s) < a \leq \varphi_e^B(k)[s] \implies a \in B_s \,\big]$$

where (as usual) B_s denotes the set B at the start of stage s. Note that $a \in B_s$ here is equivalent to $a \in B_s^{[|\tau|]}$ because $a \in \omega^{[|\tau|]}$.

The function $Restraint_\tau(s)$ is defined below.

We will need to determine whether a given computation $\Phi_e^B(k)[s]$ is σ-believable in a finite amount of time (in order to compute $\Phi_\sigma^B(k)[s]$, which is defined next, in a finite amount of time). To do this, it might seem that we need to examine each $a \in \omega^{[|\tau|]}$ for certain nodes τ. However, $\omega^{[|\tau|]}$ is an infinite set; how can we examine each of its members in a finite amount of time? The answer is that we need examine only those $a \in \omega^{[|\tau|]}$ such that $a \leq \varphi_e^B(k)[s]$.

Some intuition about σ-believability is given after the definition of $\Phi_\sigma^B(k)[s]$.

2.

$$\Phi_\sigma^B(k)[s] =_{def} \begin{cases} \Phi_e^B(k)[s], & \text{if the computation } \Phi_e^B(k)[s] \text{ is } \sigma\text{-believable} \\ \uparrow, & \text{otherwise.} \end{cases}$$

Informally, the difference between $\Phi_\sigma^B(k)[s]$ and the simpler $\Phi_e^B(k)[s]$ is as follows. Suppose $\tau \prec \sigma$, and let $i = |\tau|$.

First, consider the case $\tau^\frown \infty \preceq \sigma$; that is, σ guesses that $A^{[i]}$ is full (in other words, each member of $\omega^{[i]}$ is eventually enumerated into A). If this guess is correct then P_i requires that $B^{[i]}$ be almost full (that is, we eventually must put all but finitely many members of $\omega^{[i]}$ into B). So, we ignore the computation $\Phi_e^B(k)[s]$ in calculating the length of agreement $\ell_\sigma(s)$ (which is defined next) unless B_s already contains each $a \in \omega^{[i]}$ such that

$$Restraint_\tau(s) < a \leq \varphi_e^B(k)[s].$$

This way, node τ will not later force a number into $B^{[i]}$ that could spoil the computation $\Phi_e^B(k)[s]$.

Now consider the case $\tau^\frown f \preceq \sigma$; that is, σ guesses that $A^{[i]}$ is finite. If this guess is correct then τ can force only finitely many numbers into B, and in our proof we will pick a stage x after which it ceases to do so. This is why the definition of σ-believability does not depend on nodes τ such that $\tau^\frown f \preceq \sigma$.

The last few paragraphs provide high-level intuition; the detailed verification is in Sect. 11.5.

3.
$$\ell_\sigma(0) =_{def} 0$$
and for $s \geq 1$,

$$\ell_\sigma(s) =_{def} \begin{cases} \max\{n \leq s : (\forall k < n)[\, A_s(k) = \Phi_\sigma^B(k)[s]\,]\}, & \text{if } \sigma \in TP_s \\ \ell_\sigma(s-1), & \text{otherwise.} \end{cases}$$

Informally, $\ell_\sigma(s)$ is the minimum of s and the length of agreement between A and Φ_σ^B at the start of the greatest σ-stage less than or equal to s. As in the proof of Theorem 9, here we guarantee $\ell_\sigma(s) \leq s$ in order to ensure that $\ell_\sigma(s) < \infty$ and that each stage of the algorithm can be performed in a finite amount of time.

4.
$$restraint_\sigma(0) =_{def} 0$$
and for $s \geq 1$,

$$restraint_\sigma(s) =_{def} \begin{cases} \max\{\varphi_e^B(k)[r] : k \leq \ell_\sigma(s) \text{ and } r \leq s\}, & \text{if } \sigma \in TP_s \\ restraint_\sigma(s-1), & \text{otherwise.} \end{cases}$$

5.
$$Restraint_\sigma(0) =_{def} 0$$
and for $s \geq 1$,

$$Restraint_\sigma(s) =_{def} \begin{cases} \max\left\{ \begin{array}{l} restraint_\sigma(s), \\ \max\{Restraint_\rho(s) : \sigma \text{ is southeast of } \rho\} \end{array} \right\}, & \text{if } \sigma \in TP_s \\ Restraint_\sigma(s-1), & \text{otherwise.} \end{cases}$$

Informally, $restraint_\sigma(s)$ protects the computations, up through the length of agreement of A and Φ_σ^B at each stage $r \leq s$, whereas $Restraint_\sigma(s)$ does likewise for σ and for every stronger priority node. By "protects" we mean "tries to protect," because injuries may happen. The reader should carefully compare these definitions to the analogous ones in the proof of Theorem 9.

Even though our interest is limited to nodes on the true path, we must calculate $Restraint_\tau(s)$ during stage s for each $\tau \in TP_s$, whether or not $\tau \in TP$, because we do not know which nodes constitute the true path.

The values of ℓ_σ, $restraint_\sigma$, and $Restraint_\sigma$ can change only during σ-stages. This is quite useful; for example, if $\sigma <_L TP$ then there are only finitely many σ-stages and hence
$$\max_s Restraint_\sigma(s) < \infty.$$

Both $restraint_\sigma(s)$ and $Restraint_\sigma(s)$ are monotonic in s (see Exercise 4).

6.
$$\tilde{\ell}_\sigma =_{def} \max_s \ell_\sigma(s)$$

7.
$$\tilde{R}_\sigma =_{def} \max_s Restraint_\sigma(s)$$

We will see (in Lemma 11.2) that

$$\sigma \in TP \implies (\tilde{\ell}_\sigma < \infty \quad \text{and} \quad \tilde{R}_\sigma < \infty).$$

8. Let s be a σ-stage. Then

$$pred_\sigma(s) =_{def} \begin{cases} -1, & \text{if } s \text{ is the first } \sigma\text{-stage} \\ \max\{r < s : r \text{ is a } \sigma\text{-stage}\}, & \text{otherwise.} \end{cases}$$

9.

$$A_{-1}^{[e]} =_{def} \emptyset$$

and for $s \geq 0$,

$$A_s^{[e]} =_{def} A_s \cap \omega^{[e]}.$$

Thus, if $s \geq 0$ then $A_s^{[e]}$ is the eth row of A_s, just as $A^{[e]}$ is the eth row of A. Our sole reason for defining $A_{-1}^{[e]}$ is to simplify the condition in line 6 of Algorithm 11.1.

11.4   The Algorithm

The algorithm is as follows.

ALGORITHM 11.1

```
 1  B ← ∅.
 2  for s ← 0 to ∞
        // Compute TP_s.
 3      TP_s ← {λ}.
 4      τ ← λ.
 5      for e ← 0 to s − 1
 6          if |A_s^[e]| > |A_{pred_τ(s)}^[e]|    // Numbers have entered A^[e] since the last τ-stage.
 7              τ ← τ⌢∞
 8          else τ ← τ⌢f.

 9          TP_s ← TP_s ∪ {τ}.     // Now, |TP_s| = e + 2.

        // Now, |TP_s| = s + 1.
        // B ← B ∪ {certain elements of A}.
10      for e ← 0 to s
11          ξ ← the level e node of TP_s.
12          for each a ∈ A_s^[e] − B such that a > Restraint_ξ(s)
                // Node ξ forces a into B.
13              Put a into B.
```

11.5 Verification

As usual, B is built in stages, each of which is executable in a finite amount of time; hence B is c.e. We never put an element into B unless we know that it is in A, so $B \subseteq A$. Lemma 11.3 states that the N-requirements are met, and Lemma 11.4 does likewise for the P-requirements, thereby completing the proof of Theorem 11.

Lemma 11.1 *Let σ be the level e node of TP. Then*

$$\sigma^\frown \infty \in TP \iff A^{[e]} \text{ is full.}$$

In other words, all guesses along the true path are correct.

Proof If $\sigma^\frown \infty \in TP$ then node $\sigma^\frown \infty$ is visited infinitely often; therefore $A^{[e]}$ is infinite (verify!) and hence, since A is piecewise trivial, $A^{[e]}$ is full.

On the other hand, if $\sigma^\frown f \in TP$ then node $\sigma^\frown \infty$, being to the left of $\sigma^\frown f$, is visited only finitely often, even though σ is visited infinitely often. Therefore $A^{[e]}$ is finite and hence not full. $\qquad \square$

Lemma 11.2 contains some properties of nodes on the true path, which are used in the proofs of Lemmas 11.3 and 11.4.

Lemma 11.2 *Let σ be the level e node of TP. Then*

I. *There is a stage x such that the following four conditions are true:*

(i) $\left(\forall \tau \prec \sigma\right)\left[Restraint_\tau(x) = \tilde{R}_\tau \right]$

(ii) *No $\rho <_L \sigma$ is visited during or after stage x.*

(iii) $\left(\forall i \leq e\right)\left[A^{[i]} \text{ is finite} \implies B_x^{[i]} = B^{[i]} \right]$

(iv) $\left(\forall \sigma\text{-stage } s \geq x\right)\left(\forall k \leq \ell_\sigma(s)\right)\left[\Phi_\sigma^B(k)[s]\downarrow \implies \text{ the computation } \right.$
$\left. \Phi_e^B(k)[s] \text{ is permanent} \right]$
Note that if $k < \ell_\sigma(s)$ then $\Phi_\sigma^B(k)[s]\downarrow$, whereas it is possible that $\Phi_\sigma^B(\ell_\sigma(s))[s]\uparrow$ (as in Fig. 9.1, but not in Fig. 9.2).

II. $\tilde{\ell}_\sigma < \infty$.
III. $\tilde{R}_\sigma < \infty$.

Proof We use strong induction; assume all three parts of the lemma for each $\tau \prec \sigma$.

Proof of Part I. There is a stage x that simultaneously satisfies conditions (i), (ii), and (iii), because:

(i) Node σ has only finitely many proper ancestors. Suppose τ is one of them. Then $\tilde{R}_\tau < \infty$ (by the inductive hypothesis). Furthermore, $Restraint_\tau(s)$ is monotonic in s (see Exercise 4).

(ii)   Let $\rho <_L \sigma$. Then there exists τ such that $\tau^\frown\infty \preceq \rho$ and $\tau^\frown f \preceq \sigma$, as in
       Fig. 11.3. Because σ is in TP, so is its ancestor $\tau^\frown f$. Therefore node $\tau^\frown\infty$
       is visited only finitely often; hence, each of its descendants, including ρ, is
       visited only finitely often.
(iii) If $A^{[i]}$ is finite then so is $B^{[i]}$, because $B \subseteq A$.

We now show that conditions (i), (ii), and (iii) on the choice of x together imply
condition (iv) on the choice of x. Fix some σ-stage $s \geq x$ and some $k \leq \ell_\sigma(s)$ such
that $\Phi_\sigma^B(k)[s]\downarrow$; we will show that the computation $\Phi_e^B(k)[s]$ is permanent. Suppose
that some node ξ forces a number a into B during a stage $t \geq s$. We will show

$$a > \varphi_e^B(k)[s]. \tag{11.2}$$

Let $i = |\xi|$. Consider three cases:

Case 1. ξ is west of σ.
       That is, $\xi <_L \sigma$. This is impossible, by condition (ii) on the choice of x, because
       $t \geq s \geq x$.
Case 2. ξ is southeast of σ.
       The situation is depicted in Fig. 11.7 (where $e < i$, although another possibility
       is that $e = i$). Then

$\begin{aligned}
a &> Restraint_\xi(t) &&\text{(by line 12 of Algorithm 11.1)}\\
&\geq Restraint_\sigma(t) &&\text{(because } \xi \text{ is southeast of } \sigma \text{, and } \xi \in TP_t)\\
&\geq restraint_\sigma(t)\\
&\geq restraint_\sigma(s) &&\text{(by the monotonicity of } restraint_\sigma \text{, because } t \geq s)\\
&\geq \varphi_e^B(k)[s] &&\text{(by the definition of } restraint_\sigma \text{, because } k \leq \ell_\sigma(s) \text{ and } \sigma \in TP_s).
\end{aligned}$

Fig. 11.7 ξ is southeast of σ

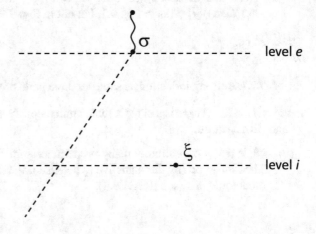

Fig. 11.8 ξ is northeast of σ

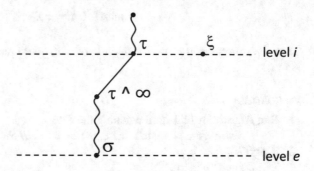

Case 3. ξ is northeast of σ.

This is where the guessing comes into play, and so it lies at the heart of this chapter. Let τ be the level i ancestor of σ, as depicted in Fig. 11.8 (in which $\tau \lhd \xi$; another possibility is that $\tau = \xi$). Note that $i < e$, because the southeast region contains all of its boundary nodes; thus $\tau \neq \sigma$. We have

$$a \in B^{[i]}_{t+1} - B^{[i]}_t \qquad \text{(because } a \text{ enters } B \text{ during stage } t)$$
$$\subseteq B^{[i]}_{t+1} - B^{[i]}_x \qquad \text{(because } t \geq s \geq x)$$
$$\subseteq B^{[i]} - B^{[i]}_x$$

and so the contrapositive of condition (iii) on the choice of x implies that $A^{[i]}$ is full. Hence (since $\sigma \in TP$ and therefore $\tau \in TP$) we have $\tau^\frown\infty \preceq \sigma$ by Lemma 11.1 (as indicated in Fig. 11.8). Since $\Phi^B_\sigma(k)[s]\downarrow$, the computation $\Phi^B_e(k)[s]$ is σ-believable; therefore, because $\tau^\frown\infty \preceq \sigma$ and $a \in \omega^{[i]}$, we have

$$\left(Restraint_\tau(s) < a \leq \varphi^B_e(k)[s] \right) \implies a \in B_s. \tag{11.3}$$

Furthermore,

$$Restraint_\tau(s) \leq Restraint_\tau(t) \qquad \text{(by the monotonicity of } Restraint_\tau \text{, because } s \leq t)$$
$$\leq Restraint_\xi(t) \qquad \text{(because } \xi \text{ is southeast of } \tau \text{, and } \xi \in TP_t)$$
$$< a \qquad \text{(because } \xi \text{ forces } a \text{ into } B \text{ during stage } t; \text{ see line 12 of Algorithm 11.1).}$$

Hence, since $Restraint_\tau(s) < a$ and $a \notin B_s$ (because a enters B during stage $t \geq s$), we have $a > \varphi^B_e(k)[s]$ by (11.3).

To summarize: Case 1 is impossible, and in either Case 2 or Case 3 we have (11.2); hence, the computation $\Phi^B_e(k)[s]$ is permanent. Therefore stage x satisfies condition (iv). This concludes the proof of Part I.

Proof of Part II. The proof of $\tilde{\ell}_\sigma < \infty$ resembles the proof of Part I of Lemma 9.2. In particular, assume for a contradiction that

$$\max_s \ell_\sigma(s) = \infty. \tag{11.4}$$

Then the following algorithm, which knows x, could determine whether a given p is in A:

ALGORITHM 11.2

1   Run Algorithm 11.1 until reaching the first
        σ-stage $s > x$ such that $\ell_\sigma(s) > p$.   // Such an s exists by (11.4).
2   **if** $\Phi_e^B(p)[s] = 1$
3   output("p is in A")
4   **else** output("p is not in A").

By condition (iv) on the choice of x, the computation $\Phi_e^B(p)[s]$ is permanent, and so

$$\Phi_e^B(p) = \Phi_e^B(p)[s]. \tag{11.5}$$

If

$$\Phi_e^B(p)[s] = 1$$

then $p \in A_s$ (because $\ell_\sigma(s) > p$), and so $p \in A$. On the other hand, if

$$\Phi_e^B(p)[s] = 0$$

then (by (11.5))

$$\Phi_e^B(p) = 0$$

and so $p \notin A$, because otherwise

$$A(p) = 1 \neq 0 = \Phi_e^B(p)$$

contradicting (11.4) (see Exercise 6). To summarize,

$$p \in A \iff \Phi_e^B(p)[s] = 1,$$

and so the output of Algorithm 11.2 is correct. Therefore A is computable, contrary to the assumption in the statement of the theorem. Thus, the assumption (11.4) has led to a contradiction, and so $\tilde{\ell}_\sigma < \infty$. This concludes the proof of Part II.

Proof of Part III. If σ is southeast of some node ρ, then

$$\rho <_L \sigma \quad \text{or} \quad \rho \prec \sigma \quad \text{or} \quad \rho = \sigma \tag{11.6}$$

(take another look at Fig. 11.6, and note that the converse need not be true). From the definitions of $\tilde{R}_\sigma$ and $Restraint_\sigma(s)$, we have

$$\tilde{R}_\sigma = \max_s \left\{ Restraint_\sigma(s) \right\}$$

$$\leq \max_s \left\{ restraint_\rho(s) : \sigma \text{ is southeast of } \rho \right\}$$

(actually, the above inequality can be shown to be an equation, but the inequality here suffices for our purposes). Therefore, by (11.6), we have

$$\tilde{R}_\sigma \leq \max_s \left(\left\{ restraint_\rho(s) : \rho <_L \sigma \right\} \cup \left\{ restraint_\tau(s) : \tau \prec \sigma \right\} \cup \left\{ restraint_\sigma(s) \right\} \right).$$
(11.7)

By condition (ii) on the choice of x,

$$\max_{s>x} \left\{ restraint_\rho(s) : \rho <_L \sigma \right\} < \infty. \tag{11.8}$$

Exercise 5, condition (i) on the choice of x, and the inductive hypothesis together imply

$$\max_{s>x} \left\{ restraint_\tau(s) : \tau \prec \sigma \right\} \leq \max_{s>x} \left\{ Restraint_\tau(s) : \tau \prec \sigma \right\} < \infty, \tag{11.9}$$

because σ has only finitely many ancestors. Part II (that is, $\tilde{\ell}_\sigma < \infty$) and condition (iv) on the choice of x together imply

$$\max_{s>x} \left\{ \varphi_\sigma^B(k)[s] : k \leq \ell_\sigma(s) \right\} < \infty,$$

and so

$$\max_{s>x} \left\{ restraint_\sigma(s) \right\} < \infty. \tag{11.10}$$

Together, (11.7)–(11.10) imply

$$\tilde{R}_\sigma < \infty.$$

QED Lemma 11.2

Lemma 11.3 N_e *is met, for all* e.

Proof This resembles the proof of Lemma 9.3, but is more complicated because it deals with ℓ_σ rather than ℓ_e.

Fix e. Let σ be the level e node of *TP*. Assume for a contradiction that N_e is not met; that is,

$$A = \Phi_e^B. \tag{11.11}$$

By Part II of Lemma 11.2, there exists

$$m = \max_s \{ \ell_\sigma(s) \}. \tag{11.12}$$

Let x be such that

(i) $A_x \upharpoonright m = A \upharpoonright m$,
(ii) $(\forall k \leq m)[$ the computation $\Phi_e^B(k)[x]$ is permanent $]$,
(iii) $(\forall \tau \prec \sigma)[Restraint_\tau(x) = \tilde{R}_\tau]$.
(iv) $m \leq x$.

Such an x exists by

(i) the A_s being an enumeration of A,
(ii) Equation (11.11) and the Permanence Lemma.
(iii) Part III of Lemma 11.2 applied to each proper ancestor of σ, and the monotonicity of $Restraint_\tau$.

Let $y > x$ be a σ-stage such that

$$A_y \upharpoonright u = A \upharpoonright u \tag{11.13}$$

where

$$u =_{def} \max\{\varphi_e^B(k)[x] : k \leq m\}.$$

By condition (ii) on the choice of x, and because $y > x$,

$$u = \max\{\varphi_e^B(k)[y] : k \leq m\}. \tag{11.14}$$

Because each ancestor of σ is in TP, Lemma 11.1 implies

$$(\forall \tau^\frown \infty \preceq \sigma)[A^{[\|\tau\|]} \text{ is full}]. \tag{11.15}$$

By (11.13) and (11.15), for each $\tau^\frown \infty \preceq \sigma$, each

$$a \in \omega^{[\|\tau\|]} - B_y$$

such that

$$Restraint_\tau(y) < a \leq u$$

is forced into B by node τ during stage y (see lines 12–13 of Algorithm 11.1), because y is a σ-stage and hence also a τ-stage. Thus,

$$(\forall \tau^\frown \infty \preceq \sigma)(\forall a \in \omega^{[\|\tau\|]})[Restraint_\tau(y) < a \leq u \implies a \in B_{y+1}]. \tag{11.16}$$

Let $z > y$ be a σ-stage. Then by condition (iii) on the choice of x, we have

$$(\forall \tau^\frown \infty \preceq \sigma)[Restraint_\tau(z) = Restraint_\tau(y) \, (= \tilde{R}_\tau)]. \tag{11.17}$$

Together, (11.14), (11.16), (11.17), condition (ii) on the choice of x, and $B_{y+1} \subseteq B_z$ imply

$$\left(\forall k \leq m\right)\left(\forall \tau ^ \infty \preceq \sigma\right)\left(\forall a \in \omega^{[|\tau|]}\right)\left[\ Restraint_\tau(z) < a \leq \varphi_e^B(k)[z] \implies a \in B_z\ \right];$$

in other words,

$$\left(\forall k \leq m\right)\left[\ \text{the compututation } \Phi_e^B(k)[z] \text{ is } \sigma\text{-believable}\ \right]. \tag{11.18}$$

For each $k \leq m$,

$$
\begin{aligned}
\Lambda_z(k) &= A(k) && \text{(by condition (i) on the choice of } x, \text{ because } z > x)\\
&= \Phi_e^B(k) && \text{(by 11.11)}\\
&= \Phi_e^B(k)[z] && \text{(by condition (ii) on the choice of } x, \text{ because } z > x)\\
&= \Phi_\sigma^B(k)[z] && \text{(by (11.18), and the definition of } \Phi_\sigma^B(k)[z]).
\end{aligned}
$$

Therefore, by condition (iv) on the choice of x (which implies $m + 1 \leq z$), and because $\sigma \in TP_z$,

$$\ell_\sigma(z) \geq m + 1$$

(take another look at the definition of $\ell_\sigma(z)$), contradicting (11.12).

$$\textit{QED Lemma } 11.3$$

Lemma 11.4 P_e is met, for all e.

Proof Fix e. Let σ be the level e node of TP. By Lemma 11.2, $\tilde{R}_\sigma < \infty$. Because σ is visited infinitely often, σ forces every member of

$$\{a \in A^{[e]} : a > \tilde{R}_\sigma\}$$

into B, unless some other node on level e forced it into B first. Thus, the set

$$A^{[e]} - B \ \left(= A^{[e]} - B^{[e]}\right)$$

is finite (in particular, $\left|\ A^{[e]} - B\ \right| \leq \tilde{R}_\sigma + 1$).

$$\textit{QED Lemma } 11.4$$

Thus, all the requirements of the theorem are met.

$$\textit{QED Theorem } 11$$

11.6   What's New in This Chapter?

Trees, of course. Let's take an informal look at the tree method as it is used in Theorem 11. Think of each tree node as a little mathematician. Collectively, the nodes on level e are responsible for meeting both P_e and N_e. They meet P_e by forcing all but finitely many of the elements of $A^{[e]}$ into B; various level e nodes might contribute to this effort. To meet N_e, on the other hand, the level e nodes work independently. Each one endeavors to meet N_e by its own length-of-agreement argument, in particular, where the length of agreement is between A and Φ_σ^B. Each one operates with its own set of e assumptions (or "guesses" as we've been calling them).

Nodes are not overworked. Consider a node σ on level e. It labors (on behalf of P_e or N_e or both) only during σ-stages. During all other stages, it naps. Its effort to meet N_e can be thwarted ("injured") by other nodes at various levels of the tree, as they force numbers into B to meet their own P-requirements. We need to protect σ's length-of-agreement argument from such injuries. Nodes southeast of σ cannot injure it, because they must respect $Restraint_\sigma(s)$. Nodes northeast of σ can indeed injure it; however, if σ's guesses are all correct (that is, $\sigma \in TP$), then they injure it only finitely often. Nodes west of σ are troublesome, because they could keep forcing numbers less than $Restraint_\sigma$ into B, while causing $Restraint_\sigma$ to grow without bound. That might indeed happen for certain nodes σ, but not for $\sigma \in TP$ (because if $\sigma \in TP$ then there are only finitely many stages s such that $TP_s <_L \sigma$). Again, what saves us is that we don't need every node on level e to succeed; rather, we need just one of them to do so. The node at level e on the true path will succeed. Some other nodes at level e, to the right of the true path, might succeed, too, although it might be difficult to prove that they do. The key is that one successful node per level is enough. However, at no stage during the construction can we know which level e node lies on the true path, so we must keep every node employed. Each node just keeps plugging away (whenever it is visited); if it does happen to lie on the true path, then its efforts will not have been in vain. *Thus, even though the entire proof is called an "infinite injury" argument, for nodes in TP it's just a finite injury argument.*

Here is another perspective, quoting [Cs], on tree-based algorithms for coping with infinite injury (the bracketed comments are mine):

> The idea of "injury" made a lot of sense for describing the original finite injury priority arguments. However, with more complicated arguments, perhaps "infinite injury" is the wrong way to think about it. Really what is happening during these constructions, is that based on a current approximation [that is, TP_s], a view is determined, and a next step is taken in a construction. The important thing is to have damage control for the harm that can be done at a stage with a wrong guess [that is, when visiting a node not in TP], and to make sure that enough good things happen when you have a good guess [that is, when visiting a node in TP].

11.7 Afternotes

1. It might seem incongruous that our Theorem 11 is called a "lemma" in the literature. That's because various (and somewhat stronger) versions of this result, and of our Theorem 12, are used to prove results about the c.e. "degrees" (see [So87] for details). Because our focus is on algorithms, we regard this result as a theorem on a par with all of our theorems in Chap. 4 and beyond.
2. Many authors (e.g., [DH]) use the phrase σ-*believability* in various tree arguments essentially the way we do. Another name for it is σ-*correctness* (Definition 3.2 of Chap. XIV of [So87]).

   Is there a useful generalization of this concept? Perhaps more than two levels of believability, defined in some way, might be helpful for certain applications.
3. Nodes that lie to the left of the true path are visited only finitely often, and hence are uninteresting. Nodes on the true path are studied extensively, because they enable us to prove that the requirements are met. To the right of the true path lies *terra incognita*. Some of those nodes might be visited infinitely often, and some not. Some of their corresponding variables (such as their length of agreement functions, or their restraint functions) might grow without bound, and some not. There is no published research about them, as far as this author is aware. Perhaps there are non-trivial results about those nodes—gems waiting to be unearthed— for this and for other tree algorithms. Perhaps, by looking at those nodes, we could categorize various tree algorithms in an interesting way. Perhaps, for example, tree algorithms in which there are only finitely many infinite paths are more limited in power than those with infinitely many infinite paths.

11.8 Exercises

1. Give a very simple proof of the Weak Thickness Lemma in the special case in which each row of the c.e.n. set A is finite.
2. Let A be piecewise trivial and c.e.n. Must $\{e : A^{[e]}$ is full$\}$ be c.e.?
3. The true path is a set of tree nodes, which we often think of as finite-length strings. Lexicographic order on these strings gives us a computable, one-to-one correspondence between them and ω. In this sense, we can think of TP as a set of natural numbers.

   (a) Give an example of a piecewise trivial, c.e.n. set A for which TP is computable.
   (b) Give an example of a piecewise trivial, c.e.n. set A for which TP is non-computable.

4. Prove that

$$(\forall \sigma)(\forall s)\big[restraint_\sigma(s) \leq restraint_\sigma(s+1) \quad \text{and} \quad Restraint_\sigma(s) \leq Restraint_\sigma(s+1)\big].$$

5. Prove that
$$(\forall \sigma)(\forall s)\big[restraint_\sigma(s) \leq Restraint_\sigma(s)\big].$$

6. Let σ be a node on level e of the tree. Prove that

$$(\exists q)[\, \Phi_e^B(q) \downarrow \quad \text{and} \quad A(q) \neq \Phi_e^B(q)\,] \implies \max_s \ell_\sigma(s) < \infty.$$

7. Suppose that we replace Φ_σ^B by Φ_e^B in the definition of ℓ_σ. In other words, we ignore the guesses represented by σ. Explain how the proof would fail.

8. During stage s, a node $\xi \in TP_s$ on a level $e < s$ might force elements into B whether

$$\xi \hat{\ } \infty \in TP_s$$

or

$$\xi \hat{\ } f \in TP_s.$$

Suppose that we modify the algorithm so that only those nodes ξ such that $\xi \hat{\ } \infty \in TP_s$ may do so. That is, suppose we replace the code

       // $B \leftarrow B \cup \{$certain elements of $A\}$.
       for $e \leftarrow 0$ **to** $s - 1$
          $\xi \leftarrow$ the level e node of TP_s.
          for each $a \in A_s^{[e]} - B$ such that $a > Restraint_\xi(s)$
             // Node ξ *forces* a into B.
             Put a into B.

by

       // $B \leftarrow B \cup \{$certain elements of $A\}$.
       for $e \leftarrow 0$ **to** $s - 1$
          $\xi \leftarrow$ the level e node of TP_s.
          for each $a \in A_s^{[e]} - B$ such that $a > Restraint_\xi(s)$
             if $\xi \hat{\ } \infty \prec TP_s$
                // Node ξ *forces* a into B.
                Put a into B.

Would the algorithm still work (that is, would the constructed set B still meet all of the requirements)?

9. During each stage we try to put certain numbers into B. What if we allowed numbers to enter B only during stages that are powers of 2? Would the algorithm still work?

10. A set S is *cofinite* if $\bar{S}$ is finite. Note that ∞ and f are the only two nodes on level 1 of the tree.

   (a) Is it possible that the set of ∞-stages is cofinite?
   (b) Is it possible that the set of f-stages is cofinite?

11. Recall the computation $\Phi_e^B(k)[s]$ is defined as σ-believable if

$$\left(\forall \tau ^\frown \infty \preceq \sigma\right)\left(\forall a \in \omega^{[|\tau|]}\right)\left[\, Restraint_\tau(s) < a \leq \varphi_e^B(k)[s] \implies a \in B_s \,\right].$$

   (a) If we change the original definition to

$$\left(\forall \tau ^\frown \infty \preceq \sigma\right)\left(\forall a \in \omega^{[|\tau|]}\right)\left[\, Restraint_\tau(s) < a \leq \varphi_e^B(k)[s] + 1000 \implies a \in B_s \,\right].$$

   would the algorithm still work?
   (b) If we change the original definition to

$$\left(\forall \tau ^\frown \infty \preceq \sigma\right)\left(\forall a \in \omega^{[|\tau|]}\right)\left[\, Restraint_\tau(s) < a \leq restraint_\sigma(s) \implies a \in B_s \,\right].$$

   would the algorithm still work?

12. In Lemma 11.2, we showed that

$$\tilde{R}_\sigma < \infty$$

   if $\sigma \in TP$.

   (a) Would it still be true if $\sigma <_L TP$?
   (b) Would it still be true if $TP <_L \sigma$?
       See the remarks about *terra incognita* in the Afternotes section.

13. In line 2 of Algorithm 11.2, suppose that we change the condition

$$\Phi_e^B(p)[s] = 1$$

   to

$$\Phi_\sigma^B(p)[s] = 1.$$

   Would that algorithm still correctly determine whether $p \in A$?

14. Define the *injury set* for a node σ as

$$I_\sigma = \left\{ a : a \text{ enters } B \text{ during a stage } s \text{ such that } a \leq Restraint_\sigma(s) \right\}.$$

If $\sigma \in TP$ then I_σ is finite. Can there be a node τ such that I_τ is infinite? If so, then what can you say about τ?

15. For each $\sigma \in TP$, there is a stage x after which no $\rho <_L \sigma$ is visited; this fact is essential in this chapter, and (as far as I know) in the verification of all algorithms based on priority trees.

   Must there be a stage after which no $\rho <_L TP$ is visited?

16. Is there a standard enumeration a_0, a_1, ... of a piecewise trivial, c.e.n. set A such that, if it were given as the input to Algorithm 11.1, the set

$$\{s : TP_s \subset TP\}$$

would be finite?

Chapter 12
An Infinitely Branching Tree (Thickness Lemma)

The priority tree used in Chap. 11 was binary, that is, each node had exactly two children. Our next tree is infinitely branching, that is, each node has infinitely many children.

A set A is *piecewise computable* if for all e, $A^{[e]}$ is computable. Because every piecewise trivial set is also piecewise computable, the following result generalizes the Weak Thickness Lemma.

Theorem 12 (Thickness Lemma) *Let A be c.e.n. and piecewise computable. Then there is a thick, c.e. subset B of A such that $A \not\leq_T B$.*

Proof The requirements are exactly the same as for the Weak Thickness Lemma:

$$P_e : A^{[e]} - B^{[e]} \text{ is finite}$$
$$N_e : A \neq \Phi_e^B$$

for each e.

12.1 The Tree

We write $S \triangle T$ to denote the *symmetric difference* between sets S and T, that is,

$$S \triangle T =_{def} (S - T) \cup (T - S).$$

For each e, because $A^{[e]}$ is computable, there exists an r such that W_r and $A^{[e]}$ are complements, that is,

$$A^{[e]} \triangle W_r = \omega,$$

© The Author(s), under exclusive license to Springer Nature Switzerland AG 2023
K. J. Supowit, *Algorithms for Constructing Computably Enumerable Sets*,
Computer Science Foundations and Applied Logic,
https://doi.org/10.1007/978-3-031-26904-2_12

Fig. 12.1 A node and its branches

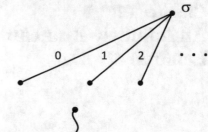

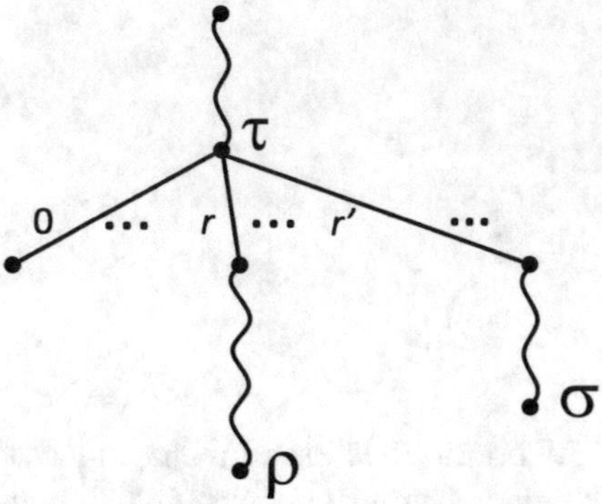

Fig. 12.2 $\rho <_L \sigma$

and so

$$A^{[e]} \bigtriangleup W_r^{[e]} = \omega^{[e]}. \tag{12.1}$$

Each node σ on level e of the tree has branches labeled with the elements of ω, in ascending order from left to right, as is depicted in Fig. 12.1.

As in the proof of the Weak Thickness Lemma, σ represents a sequence of e numbers, which we think of as a sequence of guesses. Node $\sigma ^\frown r$ represents the guess that r is the *least* number such that (12.1) is true.

With this infinitely branching tree, we write $\rho <_L \sigma$ if

$$\left(\exists \tau, r, r'\right)\left[r < r' \text{ and } \tau ^\frown r \preceq \rho \text{ and } \tau ^\frown r' \preceq \sigma \right], \tag{12.2}$$

as is illustrated in Fig. 12.2. This is a natural generalization of the definition of $\rho <_L \sigma$ that we used for the binary tree in the proof of the Weak Thickness Lemma (see Fig. 11.3).

12.2   Definitions, and a Fact

Let σ be a node on level e of the tree. Then a computation $\Phi_e^B(k)[s]$ is σ-*believable* if

$$\left(\forall \tau ^\frown r \preceq \sigma\right)\left(\forall a \in \omega^{[|\tau|]} - W_r[s]\right)\left[\text{Restraint}_\tau(s) < a \leq \varphi_e^B(k)[s] \implies a \in B_s \right]. \tag{12.3}$$

						k			
	0	1	2	3	4	5	6	7	8
$A(\langle e, k \rangle)[s]$	0	0	1	0	1	0	1	0	1
$W_r(\langle e, k \rangle)[s]$	1	1	0	1	0	1	1	0	0

Fig. 12.3 The length of disagreement is 6

Carefully compare this to the definition of σ-believability in Chap. 11.

We define $\Phi_\sigma^B(k)[s]$, $\varphi_\sigma^B(k)[s]$, $\ell_\sigma(s)$, $restraint_\sigma(s)$, $Restraint_\sigma(s)$, $\tilde{\ell}_\sigma$, and $\tilde{R}_\sigma$ exactly as in Chap. 11, except now these definitions are based on our modified definition of σ-believability.

Informally, Algorithm 12.1 uses $\Phi_\sigma^B(k)[s]$ rather than $\Phi_e^B(k)[s]$ so that it need not worry about the computation getting spoiled (Algorithm 11.1 did analogously). In particular, suppose $\tau ^\frown r \preceq \sigma$. Let $i = |\tau|$. If $\tau ^\frown r$ represents a correct guess, then

$$A^{[i]} = \omega^{[i]} - W_r$$

and so P_i requires that $B^{[i]}$ contain all but finitely many members of $\omega^{[i]} - W_r$. So, we ignore the computation $\Phi_e^B(k)[s]$ in calculating the length of agreement $\ell_\sigma(s)$ unless B_s already contains each $a \in \omega^{[i]} - W_r[s]$ such that

$$Restraint_\tau(s) < a \leq \varphi_e^B(k)[s].$$

This way, node τ will not later force a number into $B^{[i]}$ that could spoil the computation $\Phi_e^B(k)[s]$.

Recall that ℓ_σ denotes the length of agreement function between A and Φ_σ^B. We define also, for each r, the *length of disagreement* between $A^{[e]}$ and $W_r^{[e]}$ as follows:

$$\overline{\ell}_{e,r}(s) =_{def} \max \left\{ n \leq s : (\forall k < n)\big[A(\langle e, k \rangle)[s] \neq W_r(\langle e, k \rangle)[s] \big] \right\}.$$

In Fig. 12.3, $\overline{\ell}_{e,r}(s) = 6$, assuming that $s \geq 6$. Compare the length of disagreement in Fig. 12.3 with the length of agreement in Figs. 9.1 and 9.2. See Exercise 2.

A function $f : \omega \to \omega$ is *unimodal* if there is a b such that

$$(\forall a < b)\big[f(a) \leq f(a+1) \big]$$

and

$$(\forall c \geq b)\big[f(c) \geq f(c+1) \big].$$

Fact 12.1 $(\forall e, r)\left[\ \overline{\ell}_{e,r}\ \text{is unimodal or monotonically non-decreasing}\ \right].$

The proof of Fact 12.1 is left as Exercise 3.

12.3 The Algorithm

The algorithm is as follows:

ALGORITHM 12.1

```
1   B ← ∅.
2   for s ← 0 to ∞
        // Compute TPₛ.
3       TPₛ ← {λ}.
4       τ ← λ.
5       for e ← 0 to s − 1
6           if (∃j ≤ s)[ ℓ̄ₑ,ⱼ(s) > ℓ̄ₑ,ⱼ(predτ(s)) ]   // ℓ̄ₑ,ⱼ has grown since the last τ-stage.
7               r ← the least such j
8           else r ← s.

9           τ ← τ⌢r.
10          TPₛ ← TPₛ ∪ {τ}.

        // B ← B ∪ {certain elements of A}.
11      for e ← 0 to s
12          ξ ← the level e node of TPₛ.
13          for each a ∈ Aₛ^[e] − B such that a > Restraintξ(s)
                // Node ξ forces a into B.
14              Put a into B.
```


Notes on Algorithm 12.1:

1. In line 6, if s is the first τ-stage, then $pred_\tau(s) = -1$. Define $\overline{\ell}_{e,j}(-1) = -1$.
2. During each stage of Algorithm 12.1, two tasks are performed: compute TP_s, and add certain elements to B. The same is true for Algorithm 11.1.
3. Lines 11–14 of Algorithm 12.1 are identical to lines 10–13 of Algorithm 11.1.

12.4 Verification

The true path for Algorithm 12.1 is defined word-for-word as in (11.1).

All arguments based on priority trees need TP to be an infinite path. For finitely branching trees (such as the binary tree in Chap. 11), it is obvious that TP is an infinite path. It is not so obvious for infinitely branching trees. For Algorithm 12.1, it follows from Lemma 12.1. Lemmas 12.1, 12.2, 12.3, and 12.4 are analogous to Lemmas 11.1, 11.2, 11.3, and 11.4, respectively.

For each e, let

$$J_e =_{def} \{ j : A^{[e]} \triangle W_j^{[e]} = \omega^{[e]} \}.$$

Fact 12.2 For each e, J_e is non-empty.

Proof Fix e. Because A is piecewise computable, $A^{[e]}$ is computable. Therefore, letting

$$D = \omega^{[e]} - A^{[e]},$$

D is the difference of two computable sets, and hence itself is computable. Therefore D is c.e., and so it equals W_j for some j. Furthermore,

$$A^{[e]} \triangle W_j = A^{[e]} \triangle \left(\omega^{[e]} - A^{[e]} \right) = \omega^{[e]},$$

and so $j \in J_e$. Hence, J_e is non-empty. $\square$

Lemma 12.1 *Suppose that σ is on level e of TP. Let*

$$r_e = \min(J_e)$$

(this minimum exists, by Fact 12.2). Then $\sigma {}^\frown r_e \in TP$.

Proof Because $\sigma \in TP$, there is a stage y after which no $\rho \lhd \sigma$ is visited.

Assume for a contradiction that $\sigma {}^\frown j$ is visited infinitely often for some $j < r_e$. Then $\bar{\ell}_{e,j}$ is not unimodal. Therefore, by Fact 12.1, it is monotonic, and so

$$A^{[e]} \triangle W_j^{[e]} = \omega^{[e]},$$

that is, $j \in J_e$, contradicting $r_e = \min(J_e)$. Hence there is a stage $z > y$ after which no node $\tau \lhd \sigma {}^\frown r_e$ is visited.

To show $\sigma {}^\frown r_e \in TP$, it remains only to show that $\sigma {}^\frown r_e$ is visited infinitely often. Because $\sigma \in TP$, there are infinitely many σ-stages. Therefore, because $r_e \in J_e$, there are infinitely many σ-stages

$$s > z$$

such that

$$\bar{\ell}_{e,r_e}(s) > \bar{\ell}_{e,r_e}\left(pred_\sigma(s)\right);$$

each of these is a $(\sigma {}^\frown r_e)$-stage.

$$QED \ Lemma \ 12.1$$

The root of the tree is in TP. By Lemma 12.1, each node in TP has a child in TP. Therefore, by ordinary induction on the level of the tree, TP is an infinite path. Also by Lemma 12.1, each guess (as specified in Sect. 12.1) in TP is correct.

Lemma 12.2 *Let σ be the level e node of TP. Then*

I. *There is a stage x such that the following three conditions are true:*

   (i) $\left(\forall \tau \prec \sigma\right)\left[Restraint_\tau(x) = \tilde{R}_\tau \right]$
   (ii) *No $\rho <_L \sigma$ is visited during or after stage x.*
   (iii) $\left(\forall \sigma\text{-stage } s \geq x\right)\left(\forall k \leq \ell_\sigma(s)\right)\left[\Phi_\sigma^B(k)[s]\downarrow \implies \text{the computation} \right.$
       $\left. \Phi_e^B(k)[s] \text{ is permanent} \right]$

II. $\tilde{\ell}_\sigma < \infty.$
III. $\tilde{R}_\sigma < \infty.$

Note that conditions (i), (ii), and (iii) here are identical to conditions (i), (ii), and (iv) of Part I of Lemma 11.2, respectively. There is no analog here to condition (iii) of Part I of Lemma 11.2. Parts II and III here are identical to Parts II and III of Lemma 11.2, respectively.

Proof The proofs of Lemmas 11.2 and 12.2 are similar. Again, we use strong induction, assuming all three parts of the lemma for each $\tau \prec \sigma$.

Proof of Part I There is a stage x that simultaneously satisfies conditions (i) and (ii), because:

(i) Node σ has only finitely many proper ancestors. Suppose τ is one of them. Then $\tilde{R}_\tau < \infty$ (by the inductive hypothesis). Furthermore, $Restraint_\tau(s)$ is monotonic in s.
(ii) Let $\rho <_L \sigma$. Then there exist τ, r, and r' such that $r < r'$ and $\tau^\frown r \preceq \rho$ and $\tau^\frown r' \preceq \sigma$, as in Fig. 12.2. Because σ is in *TP*, so is its ancestor $\tau^\frown r'$. Therefore node $\tau^\frown r$ is visited only finitely often; hence, each of its descendants, including ρ, is visited only finitely often.

We now show that conditions (i) and (ii) on the choice of x together imply condition (iii) on the choice of x. Fix some σ-stage $s \geq x$ and some $k \leq \ell_\sigma(s)$ such that $\Phi_\sigma^B(k)[s]\downarrow$; we will show that the computation $\Phi_e^B(k)[s]$ is permanent. Suppose that some node ξ forces a number a into B during a stage $t \geq s$. We will show

$$a > \varphi_e^B(k)[s]. \tag{12.4}$$

Let $i = |\xi|$. Consider three cases:

Case 1. ξ is west of σ.
   That is, $\xi <_L \sigma$. This is impossible, by condition (ii) on the choice of x, because $t \geq s \geq x$.
Case 2. ξ is southeast of σ.
   The situation is depicted in Fig. 11.7 (where $e < i$, although another possibility is that $e = i$). Then

Fig. 12.4 ξ is northeast of σ

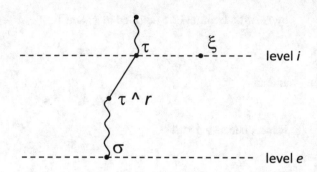

$$
\begin{aligned}
a &> Restraint_\xi(t) && \text{(by line 13 of Algorithm 12.1)} \\
&\geq Restraint_\sigma(t) && \text{(because } \xi \text{ is southeast of } \sigma, \text{ and } \xi \in TP_t) \\
&\geq restraint_\sigma(t) \\
&\geq restraint_\sigma(s) && \text{(by the monotonicity of } restraint_\sigma, \text{ because } t \geq s) \\
&\geq \varphi_e^B(k)[s] && \text{(by the definition of } restraint_\sigma, \text{ because } k \leq \ell_\sigma(s) \text{ and } \sigma \in TP_s).
\end{aligned}
$$

Case 3. ξ is northeast of σ.

Here is where the proofs of Lemmas 11.2 and 12.2 differ most.

Let τ be the level i ancestor of σ. Note that $i < e$, because the southeast region contains all of its boundary nodes; thus $\tau \neq \sigma$. Let r be such that $\tau \frown r \preceq \sigma$, as depicted in Fig. 12.4 (in which $\tau \lhd \xi$; another possibility is that $\tau = \xi$).

Since $\Phi_\sigma^B(k)[s]\downarrow$, the computation $\Phi_e^B(k)[s]$ is σ-believable; therefore, because $\tau \frown r \preceq \sigma$, by (12.3) we have

$$
\left(\forall a \in \omega^{[i]} - W_r[s]\right)\left[\, Restraint_\tau(s) < a \leq \varphi_e^B(k)[s] \implies a \in B_s \,\right].
$$

Therefore, because $a \in \omega^{[i]}$,

$$
\left(a \notin W_r[s] \ \text{ and } \ Restraint_\tau(s) < a \leq \varphi_e^B(k)[s]\right) \implies a \in B_s. \tag{12.5}
$$

Furthermore,

$$
\begin{aligned}
Restraint_\tau(s) &\leq Restraint_\tau(t) && \text{(by the monotonicity of } Restraint_\tau, \text{ because } s \leq t) \\
&\leq Restraint_\xi(t) && \text{(because } \xi \text{ is southeast of } \tau, \text{ and } \xi \in TP_t) \\
&< a && \text{(because } \xi \text{ forces } a \text{ into } B \text{ during stage } t; \text{ see line 13 of Algorithm 12.1).}
\end{aligned}
$$

Hence, since $Restraint_\tau(s) < a$ and $a \notin B_s$ (because a enters B during stage $t \geq s$), we have

$$
a \in W_r[s] \quad \text{or} \quad a > \varphi_e^B(k)[s] \tag{12.6}
$$

by (12.5).[1] Lemma 12.1 (applied to τ) and $\tau^\frown r \preceq \sigma \in TP$ together imply

$$r \in J_i$$

and so

$$A^{[i]} \triangle W_r^{[i]} = \omega^{[i]}.$$

Hence, because $a \in A^{[i]}$,

$$a \notin W_r.$$

Therefore, by (12.6), we have $a > \varphi_e^B(k)[s]$.

To summarize: Case 1 is impossible, and in either Case 2 or Case 3 we have (12.4); hence, the computation $\Phi_e^B(k)[s]$ is permanent. Therefore stage x satisfies condition (iii). This concludes the proof of Part I of Lemma 12.2.

The proofs of Parts II and III of Lemma 12.2 are word-for-word the same as those of Lemma 11.2, except that here we replace "condition (iv)" by "condition (iii)."

QED Lemma 12.2

Lemma 12.3 N_e is met, for all e.

Proof This resembles the proof of Lemma 11.3.

Fix e. Let σ be the level e node of TP. Assume for a contradiction that N_e is not met; that is,

$$A = \Phi_e^B. \tag{12.7}$$

By Part II of Lemma 12.2, there exists

$$m = \max_s \{\, \ell_\sigma(s) \,\}. \tag{12.8}$$

Let x be such that

(i) $A_x \restriction m = A \restriction m$,
(ii) $(\forall k \leq m)\big[$ the computation $\Phi_e^B(k)[x]$ is permanent$\big]$,
(iii) $(\forall \tau \prec \sigma)\big[Restraint_\tau(x) = \tilde{R}_\tau \big]$,
(iv) $m \leq x$.

Such an x exists by

(i) the A_s being an enumeration of A,
(ii) Equation (12.7) and the Permanence Lemma,
(iii) Part III of Lemma 12.2 applied to each proper ancestor of σ, and the monotonicity of $Restraint_\tau$.

---

[1] Boolean algebra is a joy forever, in my opinion.

Let $y > x$ be a σ-stage such that

$$A_y \upharpoonright u = A \upharpoonright u \tag{12.9}$$

and

$$(\forall \tau \smallfrown r \preceq \sigma)[\, W_r[y] \upharpoonright u = W_r \upharpoonright u \,] \tag{12.10}$$

where

$$u =_{def} \max \{ \varphi_e^B(k)[x] : k \le m \}.$$

By condition (ii) on the choice of x, and because $y > x$,

$$u = \max \{ \varphi_e^B(k)[y] : k \le m \}. \tag{12.11}$$

Because each ancestor of σ is in TP, Lemma 12.1 implies

$$(\forall \tau \smallfrown r \preceq \sigma)[\, A^{[|\tau|]} \triangle W_r^{[|\tau|]} = \omega^{[|\tau|]} \,]. \tag{12.12}$$

Let τ and r be such that $\tau \smallfrown r \preceq \sigma$, and for notational convenience let $i = |\tau|$. Then

$$
\begin{aligned}
A_y^{[i]} \upharpoonright u &= A^{[i]} \upharpoonright u && \text{(by (12.9))} \\
&= (\omega^{[i]} - W_r^{[i]}) \upharpoonright u && \text{(by (12.12))} \\
&= (\omega^{[i]} - W_r) \upharpoonright u \\
&= (\omega^{[i]} - W_r[y]) \upharpoonright u && \text{(by (12.10))};
\end{aligned}
$$

thus,

$$A_y^{[i]} \upharpoonright u = (\omega^{[i]} - W_r[y]) \upharpoonright u. \tag{12.13}$$

By (12.9), each

$$a \in A^{[i]} - B_y$$

such that

$$Restraint_\tau(y) < a \le u$$

is forced into B by node τ during stage y (see lines 13–14 of Algorithm 12.1), because y is a σ-stage and hence also a τ-stage. Therefore

$$(\forall a \in A^{[i]})[\, Restraint_\tau(y) < a \le u \implies a \in B_{y+1} \,]$$

and so, by (12.9) and (12.13),

$$(\forall a \in \omega^{[i]} - W_r[y])[\, Restraint_\tau(y) < a \le u \implies a \in B_{y+1} \,]. \tag{12.14}$$

Let $z > y$ be a σ-stage. Then, by condition (iii) on the choice of x, we have

$$Restraint_\tau(z) = Restraint_\tau(y)(= \tilde{R}_\tau). \qquad (12.15)$$

Together, (12.11), (12.14), (12.15), condition (ii) on the choice of x, and $B_{y+1} \subseteq B_z$ imply

$$(\forall k \leq m)(\forall \tau \smallfrown r \leq \sigma)(\forall a \in \omega^{[|\tau|]} - W_r[z])\big[\, Restraint_\tau(z) < a \leq \varphi_e^B(k)[z] \implies a \in B_z \,\big];$$

in other words,

$$(\forall k \leq m)\big[\text{the computation } \Phi_e^B(k)[z] \text{ is } \sigma\text{-believable}\big]. \qquad (12.16)$$

For each $k \leq m$,

$$
\begin{aligned}
A_z(k) = A(k) & \qquad \text{(by condition (i) on the choice of } x) \\
= \Phi_e^B(k) & \qquad \text{(by 12.7)} \\
= \Phi_e^B(k)[z] & \qquad \text{(by condition (ii) on the choice of } x) \\
= \Phi_\sigma^B(k)[z] & \qquad \text{(by (12.16), and the definition of } \Phi_\sigma^B(k)[z]).
\end{aligned}
$$

Therefore, by condition (iv) on the choice of x, and the definition of $\ell_\sigma(z)$, and because $\sigma \in TP_z$,

$$\ell_\sigma(z) > m,$$

contradicting (12.8).

QED Lemma 12.3

Lemma 12.4 P_e is met, for all e.

Proof The proof of Lemma 12.4 is word-for-word the same as that of Lemma 11.4. Thus, all the requirements of the theorem are met.

QED Theorem 12

12.5   What's New in This Chapter?

1. An infinitely branching priority tree. Can Theorem 12 be proved (without too much extra effort) by means of a binary priority tree?
2. We compute the length of *disagreement* between two functions, namely, $A^{[e]}$ and $W_r^{[e]}$. One could think of this as the length of agreement between $\omega^{[e]}$ and the set

$$A^{[e]} \bigtriangleup W_r^{[e]}.$$

3. Each of our previous length of agreement functions was used to define a restraint function; that is, it was used to decide which computations to preserve by restraining elements from entering a set B. This is likewise the case for the length of agreement function ℓ_σ in this chapter.

   The length of disagreement function $\overline{\ell}_{e,r}$ in this chapter, on the other hand, is not used to define a restraint function (A and W_r are given rather than constructed, so we cannot restrain elements from entering them); rather, $\overline{\ell}_{e,r}$ is used only to construct TP_s.

12.6 Afternotes

As we have said, in all arguments based on priority trees, TP must be an infinite path. For finitely branching trees (such as the binary tree in Chap. 11), this property of TP is obvious. For the infinitely branching tree here in this chapter, it might not be obvious but we prove it rather easily. For the infinitely branching tree in Chap. 14, we prove it with a more intricate argument, because we do not see an easy one.

12.7 Exercises

1. Let A be c.e.n. and piecewise computable, and let B be a thick, c.e. subset of A such that $A \not\leq_T B$. Must B be infinite?
2. The length of agreement (denoted $\ell_{e,i}(s)$ in Chap. 9, and $\ell_\sigma(s)$ in Chaps. 11, in this chapter, and 13) is capped at s, in order to ensure that it is finite and can be computed in a finite amount of time, as we have pointed out. On the other hand, the length of disagreement, $\overline{\ell}_{e,r}(s)$, defined in this chapter, is not explicitly capped. Must it always be finite? In other words, must, for each e, r, and s, the set

$$\left\{ n : \left(\forall k < n\right)\left[A(\langle e, k\rangle)[s] \neq W_r(\langle e, k\rangle)[s] \right] \right\}$$

   have a maximum?
3. Prove Fact 12.1.
4. Recall that $\tau^\frown r$ represents the guess that r is the least number such that (12.1) is true, where $e = |\tau|$. What if it represented the simpler guess that r is *some* number such that (12.1) is true? Then TP_s could be computed like this:

// Compute TP_s.

$\tau \leftarrow \lambda$.

for $e \leftarrow 0$ **to** $s - 1$
  if $(\exists j \leq s)[\overline{\ell}_{e,j}(s) > \overline{\ell}_{e,j}(s - 1)]$
        $r \leftarrow$ some such j
  else $r \leftarrow s$.

$\tau \leftarrow \tau^\frown r$.
$TP_s \leftarrow \tau$.

The only change we made here in Algorithm 12.1 is to replace "the least such j" by "some such j."

Would the constructed set still meet all of the requirements? Justify your answer.

5. Generalize the definition of the true path (specified in (11.1)) by defining, for each node σ,

$TP(\sigma) =_{def}$ {the leftmost descendant of σ on level e that is visited infinitely often : $e \in \omega$}.

Thus, $TP = TP(\lambda)$. For Algorithm 12.1, prove that if σ is visited infinitely often, then $TP(\sigma)$ is infinite.

6. Suppose that $\sigma^\frown r \in TP$ in Algorithm 12.1.

(a) Let j be such that $W_j \neq W_r$. Can $\sigma^\frown j$ be visited infinitely often?

(b) Let

$$J = \left\{ j : W_j \neq W_r \right\}.$$

Can the set

$$\left\{ s : (\exists j \in J)[s \text{ is a } (\sigma^\frown j)\text{-stage}] \right\}$$

be infinite?

7. Instead of the binary tree that we used in Chap. 11, we could use the infinitely branching tree depicted in Fig. 12.5. Here the branch labeled ∞ represents the guess that $A^{[|\sigma|]}$ is full; for all j, the branch labeled j represents the guess that

$$\left| A^{[|\sigma|]} \right| = j.$$

This tree holds more information about A than does the binary tree of Chap. 11.

(a) Write pseudo-code for an algorithm, based on this tree, that constructs a set B as required by Theorem 11.

Fig. 12.5 An alternative tree
for Chap. 11

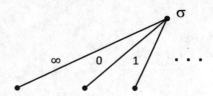

(b) Using your algorithm of part (a), let ξ be a node to the right of TP. Can there be infinitely many ξ-stages?

(c) Using your algorithm of part (a), is it true that

$$(\forall e)(\exists o \text{ on level } e)(\exists x)(\forall s > x)[\, s \text{ is a } \sigma\text{-stage}\,]?$$

Chapter 13
Joint Custody (Minimal Pair Theorem)

In Chaps. 11 and 12, the branches emanating from a tree node correspond to guesses about the given set A. In this chapter and Chap. 14, they correspond to guesses about the construction. In other words, in Chaps. 11 and 12, our tree algorithms guess about what is; in this chapter and Chap. 14, they guess about what will be.

In each of our previous length-of-agreement arguments, we showed that the length of agreement had a maximum. For example, to prove the Sacks Splitting Theorem, we proved that

$$\max_s \ell_{e,i}(s) < \infty$$

to help show that requirement $R_{e,i}$ is met. Similarly, for both of the thickness lemmas, we proved that

$$\max_s \ell_\sigma(s) < \infty$$

for each $\sigma \in TP$, to help show that N_e is met. In the current chapter, there might be some nodes $\sigma \in TP$ such that

$$\max_s \ell_\sigma(s) = \infty;$$

nevertheless, we will be able to meet our requirements.

This chapter contains a few other innovations, including the "joint custody" technique.

© The Author(s), under exclusive license to Springer Nature Switzerland AG 2023   113
K. J. Supowit, *Algorithms for Constructing Computably Enumerable Sets*,
Computer Science Foundations and Applied Logic,
https://doi.org/10.1007/978-3-031-26904-2_13

13.1   The Theorem

A *minimal pair* is a pair of c.e.n. sets A and B such that

$$(\forall \text{ c.e. } C)\big[(C \leq_T A \text{ and } C \leq_T B) \implies C \text{ is computable}\big].$$

Informally, c.e.n. sets A and B constitute a minimal pair if no c.e.n. set lies below each of them.

Theorem 13 (Lachlan-Yates) *There exists a minimal pair.*

Proof We construct A and B in stages, so as to satisfy the following requirements for each e:

$$P_{e,A} : W_e \neq \bar{A}$$
$$P_{e,B} : W_e \neq \overline{B}$$
$$N_e : \neg Q(e) \text{ or } (\Phi_e^A \text{ is a total, computable function})$$

where the predicate $Q(e)$ is true if and only if

$$\Phi_e^A = \Phi_e^B = \text{a total function}.$$

As usual, the P (or "positive") requirements are so named because they try to insert various elements into A and B. The N (or "negative") requirements try to keep certain elements out of A and B.

Lemma 13.1 *These requirements cumulatively suffice to prove the theorem.*

Proof [1]Assume that all of the requirements are met.

The $P_{e,A}$ requirements cumulatively imply the non-computability of A. The $P_{e,B}$ requirements do likewise for B. The $P_{e,A}$ and $P_{e,B}$ requirements are just like the $R_{e,i}$ requirements of Chap. 4.

Let C be a c.e. set such that $C \leq_T A$ and $C \leq_T B$. We will show that C is computable.

First, we claim that $A \neq B$. To see this, assume for a contradiction that $A = B$. Let e' be such that for all D and k,

$$\Phi_{e'}^D(k) =_{def} \begin{cases} 1, & \text{if } k \in D \\ 0, & \text{otherwise.} \end{cases}$$

Then for all D,

$$\Phi_{e'}^D = D.$$

---

[1] This is an algorithms book. It is primarily about algorithms for constructing c.e. sets to meet various collections of requirements. Hence, the reader may skip the proof of this lemma without loss of continuity.

In particular,

$$\Phi_{e'}^A = A = B = \Phi_{e'}^B \tag{13.1}$$

which is a total function; thus, $Q(e') = \text{TRUE}$. Therefore requirement $N_{e'}$, together with (13.1), implies that A is computable, a contradiction. Thus, $A \neq B$.

So, either $A - B \neq \emptyset$ or $B - A \neq \emptyset$; without loss of generality, assume the latter. Fix some $b \in B - A$. Because $C \leq_T A$ there is an i such that

$$C = \Phi_i^A,$$

and because $C \leq_T B$ there is a j such that

$$C = \Phi_j^B.$$

Thus,

$$\Phi_i^A = \Phi_j^B = C. \tag{13.2}$$

There exists e such that for all D and k,

$$\Phi_e^D(k) =_{def} \begin{cases} \Phi_i^D(k), & \text{if } b \notin D \\ \Phi_j^D(k), & \text{otherwise} \end{cases}$$

as the reader should verify. Because $b \in B - A$, we have

$$\Phi_e^A = \Phi_i^A$$

and

$$\Phi_e^B = \Phi_j^B$$

and therefore, by (13.2), we have

$$\Phi_e^A = \Phi_e^B = C$$

which is a total function. Hence $Q(e) = \text{TRUE}$, and so N_e implies that C is computable.

Therefore A and B constitute a minimal pair.   □

13.2 The Tree, and an Overview of Our Strategy

We use a binary priority tree. Emanating from each node are branches labeled ∞ and f, respectively, exactly as in Fig. 11.2. The interpretation of the guesses for this chapter is given after the definitions section.

Each node σ on level e is trying to meet three requirements: $P_{e,A}$, $P_{e,B}$, and N_e.

To meet $P_{e,A}$, node σ chooses a fresh witness $n_{\sigma,A}$ for it. Then at least one of the following events occurs:

(1) While visiting σ, we learn that $n_{\sigma,A} \in W_e$. Then we put $n_{\sigma,A}$ into A; thus, $n_{\sigma,A} \in A \cap W_e$. In this case, $n_{\sigma,A}$ is called a *positive witness* for $P_{e,A}$.

(2) While visiting some other node ξ on level e of the tree, we put $n_{\xi,A}$ into A as a positive witness for $P_{e,A}$, thereby relieving σ of its responsibility in this regard.

(3) It turns out that $n_{\sigma,A} \notin W_e$, and so we never put $n_{\sigma,A}$ into A; thus, $n_{\sigma,A} \in \overline{A} \cap \overline{W_e}$. In this case, $n_{\sigma,A}$ is called a *negative witness* for $P_{e,A}$.

In each case, $P_{e,A}$ is met. Our plan for meeting $P_{e,B}$ is analogous, substituting B for A.

To meet N_e, node σ uses a length-of-agreement argument.

The details come next.

13.3   Definitions

As in all of our tree-based algorithms, during each stage s we compute a path from the root of the tree down to a level s node that we call TP_s.

Let σ be a node on level e of the tree.

1. Let $n_{\sigma,A}$ denote the current witness for $P_{e,A}$ that is attached to node σ. Think of $n_{\sigma,A}$ as a variable; its value changes each time that σ is initialized. Likewise, $n_{\sigma,B}$ denotes the current witness for $P_{e,B}$ that is attached to σ.

2. The requirement $P_{e,A}$ *needs attention* at stage s if

$$W_e[s] \cap A_s = \emptyset \quad \text{and} \quad n_{\sigma,A} \in W_e[s]$$

(recall that the notation $W_e[y]$ is defined in (4.1)). Likewise, $P_{e,B}$ *needs attention* at stage s if

$$W_e[s] \cap B_s = \emptyset \quad \text{and} \quad n_{\sigma,B} \in W_e[s].$$

3.

$$\ell_\sigma(s) =_{def} \begin{cases} \max\left\{ n \le s : (\forall k < n)[\ \Phi_e^A(k)[s] \downarrow = \Phi_e^B(k)[s]\ ] \right\}, & \text{if } \sigma \in TP_s \\ -1, & \text{otherwise} \end{cases}$$

where $\Phi_e^A(k)[s] \downarrow = \Phi_e^B(k)[s]$ means that both are defined, and that they are equal. Thus, $\ell_\sigma(s)$ is the length of agreement between $\Phi_e^A(k)[s]$ and $\Phi_e^B(k)[s]$ (where both being undefined is regarded as a type of disagreement). As in Chaps. 9, 11, and 12, this length of agreement is capped at s to ensure that it is finite and that each stage can be computed in a finite amount of time.

4. $L_\sigma(s) =_{def} \max\{\ell_\sigma(r) : r \le s\}$.

Thus, $L_\sigma(s)$ is the "high water mark" of $\ell_\sigma(s)$ so far. Note that $L_\sigma(s)$ is monotonic as a function of s, whereas $\ell_\sigma(s)$ might not be.

5. A stage $s \geq 1$ is *σ-expansionary* if $L_\sigma(s) > L_\sigma(s-1)$. See Exercise 1.
6. To *initialize* a node σ is to assign a fresh value to $n_{\sigma,A}$ and a fresh value to $n_{\sigma,B}$. As usual, by "fresh" here we mean larger than any witness or any φ value seen so far in the construction.

Conspicuous by its absence here is some kind of restraint function, which helped us avoid certain injuries in Chaps. 9, 11, and 12. Rather, in this chapter we avoid certain injuries by picking fresh witnesses, as we did in Chaps. 7 and 8.

13.4 Interpretation of the Guesses

The ∞ branch emanating from σ represents the guess that there are infinitely many σ-expansionary stages. The f branch represents the guess that there are only finitely many such stages.

13.5 The Algorithm

ALGORITHM 13.1

```
 1  A ← ∅.
 2  B ← ∅.
 3  for s ← 0 to ∞
        // Compute TPₛ.
 4      TPₛ ← {λ}.
 5      τ ← λ.
 6      for e ← 0 to s − 1
 7          if s is a τ-expansionary stage
 8              τ ← τ⌢∞
 9          else τ ← τ⌢f.

10          TPₛ ← TPₛ ∪ {τ}.

11      Initialize each τ ∈ TPₛ that has never been initialized.
12      Initialize each node to the right of TPₛ.

        // Perhaps put a number into A or B.
13      if P_{|ξ|, A} or P_{|ξ|, B} needs attention for some ξ ∈ TPₛ
14          ξ ← the shortest such node, that is, the one closest to the root.
            // Node ξ acts.
15          if P_{|ξ|, A} needs attention
16              A ← A ∪ {n_{ξ, A}}
17          else B ← B ∪ {n_{ξ, B}}.
```

Note that during a given stage, one element may be added to A, or one to B, or to neither, *but not to both*. This observation is crucial for the "custody" argument in the proof of Lemma 13.3.

13.6 Verification

The true path TP is defined word-for-word as in (11.1).

Fact 13.1 Let σ be the level e node of TP. Then

$$Q(e) \implies \sigma^\frown\infty \in TP.$$

Proof Assume $Q(e)$, that is,

$$\Phi_e^A = \Phi_e^B = \text{a total function.} \tag{13.3}$$

Assume for a contradiciton that $\sigma^\frown f \in TP$. Then there is a stage z during or after which $\sigma^\frown\infty$ is never visited. Let $m = L_\sigma(z)$. Then, because L_σ is monotonic,

$$m = \max_s L_\sigma(s). \tag{13.4}$$

By (13.3) and the Permanence Lemma,

$$(\forall k)(\exists s_k)\big[\text{the computations } \Phi_e^A(k)[s_k] \text{ and } \Phi_e^B(k)[s_k] \text{ are both permanent}\big].$$

Because σ is on the true path and hence is visited infinitely often, there exists a σ-stage t such that

$$t > \max\{s_k : k \leq m\}.$$

Therefore, by (13.3),

$$(\forall k \leq m)\big[\ \Phi_e^A(k)[t] \downarrow = \Phi_e^A(k)[t]\ \big]$$

and so

$$L_\sigma(t) \geq m + 1,$$

contradicting (13.4).   □

As in all of our priority arguments, we define a stage after which various finitely occurring events no longer occur. In particular, for each $\sigma \in TP$, let $x(\sigma)$ denote a stage during or after which

(i) no $\rho <_L \sigma$ is visited,
(ii) no $\tau \preceq \sigma$ acts,
(iii) σ is not initialized.

Such an $x(\sigma)$ exists because

(i) $\sigma \in TP$.
(ii) No node acts more than twice: once to put an element into A, and once into B (*acting* is defined in a comment just before line 15 of Algorithm 13.1). Furthermore, σ has only finitely many ancestors.
(iii) Node σ is initialized either by line 11 of Algorithm 13.1 (which happens only once), or by line 12 when some $\rho <_L \sigma$ is visited (which happens only finitely often, because $\sigma \in TP$).

Lemma 13.2 *Each P-requirement is met.*

Proof Fix e. Let σ be the level e node of TP, and let $x = x(\sigma)$. Consider requirement $P_{e,A}$. By condition (iii) on the choice of x, node σ is never initialized during a stage $s \geq x$. Therefore, $n_{\sigma,A}(x)$ is the final value of the variable $n_{\sigma,A}$; for notational convenience, let $n = n_{\sigma,A}(x)$.

Case 1. $n \in W_e$.
Let $y \geq x$ be a σ-stage such that

$$n \in W_e[y].$$

Such a y exists because $\sigma \in TP$. Requirement $P_{e,A}$ does not need attention at stage y, because otherwise node σ would act during stage y, contradicting condition (ii) on the choice of x. Therefore

$$W_e[y] \cap A_y \neq \emptyset$$

and so $P_{e,A}$ is met.
Case 2. $n \notin W_e$.
Then n is never put into A during a visit to σ. Nor can that happen during a visit to some other node, because the witnesses chosen by procedure *Initialize* are all distinct. Therefore n is a negative witness for $P_{e,A}$.

In either case, $P_{e,A}$ is met. Analogous reasoning establishes that $P_{e,B}$ is met. $\square$

Lemma 13.3 *Each N-requirement is met.*

Proof Fix e; we will show that N_e is met. If $Q(e) = $ FALSE then the proof is immediate; so assume that $Q(e) = $ TRUE, that is,

$$\Phi_e^A = \Phi_e^B = \text{a total function.} \tag{13.5}$$

Let σ be the level e node of TP. We will show that Φ_e^A is computable, thereby implying that N_e is met.

Let

$$x = x(\sigma^\frown \infty),$$

which is defined because $\sigma^\frown \infty \in TP$ (by Fact 13.1). The following algorithm, which knows x, is given p as input. We will show that it outputs $\Phi_e^A(p)$.

ALGORITHM 13.2

1   Run Algorithm 13.1 until reaching the first $(\sigma^\frown \infty)$-stage $t_0 > x$ such that $L_\sigma(t_0) > p$.
2   Output($\Phi_e^A(p)[t_0]$).

Because $\sigma^\frown \infty$ is on the true path and hence is visited infinitely often, and each of these visits occurs during a σ-expansionary stage, the function L_σ has no maximum. Therefore the stage t_0 will indeed be found by Algorithm 13.2.

We will show that

$$\Phi_e^A(p) = \Phi_e^A(p)[t_0]$$

by means of what we call the "custody" idea.

In particular, let

$$t_0 < t_1 < \cdots$$

denote all of the $(\sigma^\frown \infty)$-stages greater than or equal to t_0. Then

$$p < \ell_\sigma(t_0) < \ell_\sigma(t_1) < \ell_\sigma(t_2) < \cdots$$

and

$$(\forall i)\big[\, \Phi_e^A(p)[t_i] = \Phi_e^B(p)[t_i] \,\big]. \tag{13.6}$$

For each z, if some $a \leq \varphi_e^A(p)[z]$ enters A during stage z then we say that the *A-side is injured during stage z*. Such an injury is caused by the action of a node ξ in line 16 of Algorithm 13.1. Likewise, if some $a \leq \varphi_e^B(p)[z]$ enters B during stage z then we say that the *B-side is injured during stage z*. Such an injury is caused by the action of a node ξ in line 17 of Algorithm 13.1.

Proposition 1 *Let $i \in \omega$, and suppose that the A-side is not injured during stage t_i. Then for each $z \in (t_i,\ t_{i+1})$, the A-side is not injured during stage z. Informally, this implies that A maintains custody of the value $\Phi_e^A(p)[t_i]$ (which, by (13.6), equals $\Phi_e^B(p)[t_i]$) from the start of stage t_i to the start of stage t_{i+1}.*

Proof Assume for a contradiction that the A-side were injured during a stage in $(t_i,\ t_{i+1})$; let z be the first such stage. Then

$$\varphi_e^A(p)[t_i] = \varphi_e^A(p)[z]. \tag{13.7}$$

Let ξ be the unique node that acts during stage z. Consider the possible relative positions of nodes $\sigma^\frown\infty$ and ξ.

Case 1. $\xi <_L \sigma^\frown\infty$.
This contradicts condition (i) on the choice of x, because $z > t_i \geq t_0 > x = x(\sigma^\frown\infty)$.

Case 2. $\sigma^\frown\infty <_L \xi$.
Then during stage t_i, which is a $(\sigma^\frown\infty)$-stage, the variable $n_{\xi,A}$ receives a fresh value (by line 12 of Algorithm 13.1) that is larger than any number seen in the construction up to that point (in particular, larger than $\varphi_e^A(p)[t_i]$). Therefore, by (13.7), the action of ξ does not cause an A-side injury during stage z, a contradiction.

Case 3. $\xi \preceq \sigma^\frown\infty$.
This contradicts condition (ii) on the choice of x, because $z > t_i \geq t_0 > x = x(\sigma^\frown\infty)$.

Case 4. $\sigma^\frown\infty \prec \xi$.
Node ξ acts during stage z, and so z is a ξ-stage and hence is also a $(\sigma^\frown\infty)$-stage. However, there is no $(\sigma^\frown\infty)$-stage in (t_i, t_{i+1}), a contradiction. $\square$

Thus, each of the four cases leads to a contradiction.

Proposition 2 *Let $i \in \omega$, and suppose that the B-side is not injured during stage t_i. Then for each $z \in (t_i, t_{i+1})$, the B-side is not injured during stage z.*

Proof Identical to the proof of Proposition 1, substituting B for A. $\square$

Proof There is no stage during which both the A-side and the B-side are injured, because at most one number enters $A \cup B$ during each stage. Therefore, for each i, either the A-side or the B-side is not injured during stage t_i; a side that is not injured maintains custody of the value $\Phi_e^A(p)[t_i]$ $(=\Phi_e^B(p)[t_i])$ from the start of stage t_i to the start of stage t_{i+1}, by Propositions 1 and 2.[2] Therefore, by (13.6) and a simple induction, we have

$$\Phi_e^A(p)[t_0] = \Phi_e^B(p)[t_0] = \Phi_e^A(p)[t_1] = \Phi_e^B(p)[t_1] = \Phi_e^A(p)[t_2] = \Phi_e^B(p)[t_2] = \cdots \tag{13.8}$$

By (13.5) we have $\Phi_e^A(p) \downarrow$; hence, by the Permanence Lemma, there is a stage y such that the computation $\Phi_e^A(p)[y]$ is permanent. Let j be such that $t_j > y$. Then the computation $\Phi_e^A(p)[t_j]$ is permanent, and so

$$\Phi_e^A(p) = \Phi_e^A(p)[t_j].$$

---

[2] It might be that neither the A-side nor the B-side is injured during stage t_i. In that case, both sides would maintain custody of the value $\Phi_e^A(p)[t_i]$ from the start of stage t_i to the start of stage t_{i+1}.

Therefore, by (13.8), we have

$$\Phi_e^A(p) = \Phi_e^A(p)[t_0].$$

Thus, Algorithm 13.2 correctly computes $\Phi_e^A(p)$. Hence Φ_e^A is computable, and so requirement N_e is met.

> *QED Lemma 13.3*

Proof The theorem follows from the three lemmas.

> *QED Theorem 13*

13.7   What's New in This Chapter?

I vacillated when naming this chapter. Most chapters in this book are named after the most salient new algorithmic technique introduced therein. What's the most interesting new technique in this chapter?

Perhaps a better title would be "Guessing about the construction," which we mentioned in the first paragraph of the chapter. Because we are guessing about what will be rather than what is, this may be viewed is a more abstract or advanced tree argument than those with which we proved the thickness lemmas. On the other hand, one might view this as a more primitive tree argument than those, because it uses no σ-specific computation function (in other words, it uses no concept of σ-believability).

A second contender for the title was "Potentially infinite length of agreement," as discussed in the second paragraph of the chapter.

A third possibility was "Preserving the result of a computation without necessarily preserving its usage." We did this preservation by using "joint custody," which is a snappier title.

13.8   Afternotes

1. The Minimal Pair Theorem was originally proved independently by Lachlan [La66] and Yates [Ya66a]. A proof using trees appears in [DH]; our proof is essentially based on that one.
2. The key technique in the proof of Lemma 13.1 is called "Posner's Remark."
3. In [Mi] there is a proof of the minimal pair theorem that uses an infinitely branching tree. Often in tree arguments, it's a matter of style as to whether to put more or less information in the tree.

4. Our definition of minimal pairs is restricted to c.e. sets. However, in the literature, minimal pairs are usually defined for sets in general.

13.9 Exercises

1. Suppose that stage s is σ-expansionary. Prove that s is a σ-stage.
2. Prove that $x(\sigma) \geq |\sigma|$, for each $\sigma \in TP$.
3. In Algorithm 13.1, to initialize a node σ on level e of the tree is to assign fresh values to $n_{\sigma,A}$ and to $n_{\sigma,B}$. Suppose that we change the initialization procedure, so that it first tests whether there already is an element in $A_s \cap W_e[s]$ and if so, it doesn't bother to change $n_{\sigma,A}$. Would the algorithm still work (that is, would all of the requirements still be met)?
4. Suppose that the initialization routine for Algorithm 13.1 were altered so that

$$n_{\sigma,A} = n_{\tau,A}$$

whenever $|\sigma| = |\tau|$. Which argument in the verification section would become invalid?
5. Prove that
$$(\forall \tau)\Big[\max_s n_{\tau,A}(s) = \infty \iff TP <_L \tau \Big].$$

6. In Algorithm 13.2, suppose that we replace

$$L_\sigma(t_0) > p$$

by

$$\ell_\sigma(t_0) > p.$$

Would the algorithm still correctly compute $\Phi_e^A(p)$?
7. Suppose that σ is the level e node of TP. Fact 13.1 says that

$$Q(e) \implies \sigma^\frown \infty \in TP.$$

Must the converse hold, that is, must it be that

$$\sigma^\frown \infty \in TP \implies Q(e)?$$

8. In the proof of Lemma 13.3, we defined $x = x(\sigma^\frown \infty)$. Suppose that instead we define $x = x(\sigma)$. Where would the proof fail?

9. In the proof of Lemma 13.3, some aspects of the computation $\Phi_e^A(p)[t_{i+1}]$ might differ from those of the computation $\Phi_e^A(p)[t_i]$. For example, perhaps

$$\varphi_e^A(p)[t_{i+1}] \neq \varphi_e^A(p)[t_i]. \tag{13.9}$$

We have proved only that their output is the same, that is,

$$\Phi_e^A(p)[t_{i+1}] = \Phi_e^A(p)[t_i].$$

Prove that (13.9) is true for only finitely many values of i.

Chapter 14
Witness Lists (Density Theorem)

The algorithm in this chapter uses and extends many of the techniques that we have seen so far, and introduces a few more.

Theorem 14 (Sacks Density Theorem) *Let A and C be c.e., and assume that $A <_T C$. Then there exists a c.e. set B such that $A <_T B <_T C$.*

Proof We prove a slightly stronger result, namely, that there exist incomparable c.e. sets B_0 and B_1 such that

$$A <_T B_0 <_T C \quad \text{and} \quad A <_T B_1 <_T C. \tag{14.1}$$

To do this, we construct c.e. sets B_0 and B_1 such that

$$B_0 \not\leq_T B_1 \quad \text{and} \quad B_1 \not\leq_T B_0 \quad \text{and} \quad A \leq_T B_0 \leq_T C \quad \text{and} \quad A \leq_T B_1 \leq_T C. \tag{14.2}$$

See Exercise 1.

Assume that we are given a standard enumeration $a_0, a_1, \ldots$ of A, and a standard enumeration $c_0, c_1, \ldots$ of C. As usual, for each s we let

$$A_s =_{def} \{a_0, a_1, \ldots, a_s\}.$$

We obtain $B_0 \not\leq_T B_1$ and $B_1 \not\leq_T B_0$ by meeting the following requirements (familiar to us from the Friedberg-Muchnik construction):

$$R_{2e} : B_1 \neq \Phi_e^{B_0}$$
$$R_{2e+1} : B_0 \neq \Phi_e^{B_1}$$

for each e.

© The Author(s), under exclusive license to Springer Nature Switzerland AG 2023
K. J. Supowit, *Algorithms for Constructing Computably Enumerable Sets*,
Computer Science Foundations and Applied Logic,
https://doi.org/10.1007/978-3-031-26904-2_14

We obtain $A \leq_T B_0$ and $A \leq_T B_1$ by "coding," that is, we "encode" A into B_0 and into B_1. In particular, during stage s, we simply put $2a_s$ into B_0 and into B_1 (see Exercise 2). All other numbers in $B_0 \cup B_1$ will be odd; thus, all numbers that we choose as witnesses to the R-requirements will be odd.

We obtain $B_0 \leq_T C$ and $B_1 \leq_T C$ by C-permitting: before an odd number can enter B_0 or B_1, it needs permission by C. In this chapter, we cannot use the most straightforward variety of permitting (that is, the one that we saw in Chap. 8) because for each $a \in A$, we must put $2a$ into B_0 and into B_1, whether or not $2a$ receives C-permission. Instead, our C-permitting uses "witness lists," which are defined in Sect. 14.2.

Thus, our overall strategy is three-pronged: we use the Friedberg-Muchnik technique to meet the R-requirements, combined with the coding of A into B_0 and B_1, and the C-permitting. The coding of A can cause infinite injury to the R-requirements; we use a priority tree to cope with that.

14.1   The Tree

Each node σ of the tree has infinitely many branches, labeled $0, 1, \ldots, R$, from left to right, as is depicted in Fig. 14.1. The branches emanating from σ represent guesses; we will specify what each branch guesses in Sect. 14.4.2 ("Interpretation of the guesses").

We write $\rho <_L \sigma$ if

$$(\exists \tau, \alpha, \beta)\big[\, \alpha \text{ and } \beta \text{ are children of } \tau, \text{ and } \alpha \lhd \beta, \text{ and } \alpha \preceq \rho, \text{ and } \beta \preceq \sigma \,\big],$$

as is illustrated in Fig. 14.2. This is analogous to the definition in (12.2).

14.2   Definitions

Throughout this section, let σ be a node on level $2e$ of the tree (if σ were on level $2e + 1$ then we would swap B_0 with B_1 in these definitions). As usual, whenever we write a variable name with an attached s, we refer to its value at the start of stage s.

Fig. 14.1 A node and its branches

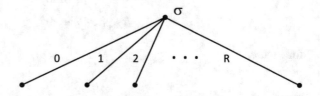

Fig. 14.2 $\rho <_L \sigma$

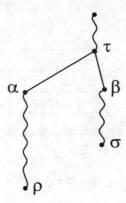

1. The *witness list* for σ is a vector n_σ. For each i and s, either $n_\sigma(i)[s]$ is a number or it is undefined. We write $n_\sigma(i)[s] \downarrow$ in the former case, and $n_\sigma(i)[s] \uparrow$ in the latter. If $n_\sigma(i)[s] \downarrow$ then we call $n_\sigma(i)[s]$ the ith *witness of* σ at the start of stage s.
Let

$$m_\sigma(s) =_{def} \begin{cases} \max I, & \text{if } I \neq \emptyset \\ -1, & \text{otherwise} \end{cases}$$

where

$$I = \{i \,:\, n_\sigma(i)[s] \downarrow\}.$$

A witness $n_\sigma(i)$ is *realized* if $i < m_\sigma$. If $i = m_\sigma$ then $n_\sigma(i)$ is *unrealized*. The variables $n_\sigma(i)$ and m_σ can change during a stage (not just at the start of a stage). Likewise, a witness can become realized, unrealized, or undefined during a stage. At all times σ has exactly m_σ realized witnesses, namely,

$$n_\sigma(0), \, n_\sigma(1), \, \ldots, \, n_\sigma(m_\sigma - 1),$$

and exactly one unrealized witness, namely, $n_\sigma(m_\sigma)$. The distinction between the adjectives "unrealized" and "undefined," as applied to witnesses, is important. The situation is illustrated in Fig. 14.3 (much information is contained in this figure; the reader should keep looking at it while studying the rest of this chapter).

i

	0	1	2	3	4	5	6 7 8 $\cdots$
$n_\sigma(i)[s]$	5	17	41	101	365	601	undefined
realized?	yes	yes	yes	yes	yes	no	not applicable
$\Phi_e^{B_0}(n_\sigma(i))[s]$	0	0	0	0	0	anything	not applicable

Fig. 14.3 The witness list for σ at the start of stage s. Here $m_\sigma(s) = 5$

Note in Fig. 14.3 that

$$n_\sigma(0) < n_\sigma(1) < \cdots < n_\sigma(m_\sigma);$$

this is always true, because witnesses are chosen as fresh numbers. Also, for each i, if $n_\sigma(i)$ is defined then it is odd.

2. A witness $b = n_\sigma(i)[s]$ is a *star witness at stage* s if it is realized (that is, $i < m_\sigma(s)$) and $b \in B_1[s]$.

3. A star witness $b = n_\sigma(i)[s]$ at stage s is *permanent* if the computation $\Phi_e^{B_0}(b)[s]$ is permanent.

If σ has a permanent star witness b then

$$B_1(b) = 1 \neq 0 = \Phi_e^{B_0}(b),$$

and so R_{2e} is met (see Exercise 3).

Furthermore, as will be seen in line 12 of the main code of the algorithm, if σ has a permanent star witness, then m_σ has an upper bound. This will be crucial in the proof of Fact 5.

4. To *cancel* σ is to undefine $n_\tau(i)$ for each τ such that $\sigma \preceq \tau$ and each $i \leq m_\tau$, and then to set m_τ to -1. In other words, to cancel a node is to discard all witnesses of that node and of its descendants.

5. For each i and s,

$$inserted(\sigma, i)[s] =_{def} \begin{cases} \text{TRUE,} & \text{if } (\exists z \leq s)\big[\, n_\sigma(i)[z] \in B_1(s) \,\big] \\ \text{FALSE,} & \text{otherwise.} \end{cases}$$

Thus, the Boolean variable $inserted(\sigma, i)$ initially equals FALSE, and stays that way until we reach a stage t during which $n_\sigma(i)[t]$ is inserted into B_1; it then equals TRUE at the start of stage $t + 1$ and forever after. If no such t exists then $inserted(\sigma, i)$ always equals FALSE.

6. For each i and s,

$$lastRealized(\sigma, i)[s] =_{def} \begin{cases} \max Z, & \text{if } i < m_\sigma(s) \\ \infty, & \text{otherwise} \end{cases}$$

where
$$Z = \{ z < s : n_\sigma(i) \text{ became realized during stage } z \}.$$

In other words, if $n_\sigma(i)$ is realized at the start of stage s, then $lastRealized(\sigma, i)[s]$ is the last stage before s during which $n_\sigma(i)$ became realized. Thus, the following three statements are equivalent:

(a) $lastRealized(\sigma, i)[s] < \infty$,
(b) $i < m_\sigma(s)$,
(c) $n_\sigma(i)$ is realized at the start of stage s.

7. For $i < m_\sigma(s)$,
$$\varphi(\sigma, i)[s] =_{def} \varphi_e^{B_0}(n_\sigma(i))[r],$$

where $r = lastRealized(\sigma, i)[s]$. Thus, during stage s, only the entrance of some $b \leq \varphi(\sigma, i)[s]$ into B_0 can spoil the computation associated with a realized witness $n_\sigma(i)[s]$.

Note that $inserted(\sigma, i)$, $lastRealized(\sigma, i)$, and $\varphi(\sigma, i)$ are not variables of the algorithm (that is, the algorithm does not explicitly assign values to them); rather, they are functions defined in terms of such variables.

8. For each i,

$$entryIntoC(i) =_{def} \begin{cases} s \text{ such that } i = c_s, & \text{if } i \in C \\ -1, & \text{otherwise.} \end{cases}$$

Note that $entryIntoC(i)$ depends neither on the current stage nor on any other variable of the algorithm. Rather, it depends only on the standard enumeration of C that we are given.

9. A number is *fresh* if it is odd and larger than any witness or usage seen during the execution of the algorithm thus far. This is just like our original definition of "fresh" from Chap. 7, except that it now requires oddness.

14.3   The Algorithm

For each σ, all witnesses of σ are initially undefined; in other words, $m_\sigma(0) = -1$.

Each stage executes a **switch** statement, which is a control structure that we have not yet used in our pseudo-code. Our **switch** statement (along with its associated keywords **case**, **break**, and **default**) works as follows: If the condition for Case 0, 1, 2, or 3 applies, then we perform the action associated with the lowest-numbered one that does. If none of them apply, then we perform the **default** action. Thus, we perform exactly one of the five actions.[1] A **break** statement causes the flow of control to exit the **switch** block. Thus, in the main code of Algorithm 14.1, whenever a **break** statement is executed, line 14 is executed next.

Thus, we could replace the **switch** statement in the main code by the following, equivalent code (assuming that we have filled in the conditions for Cases 0, 1, 2, and 3):

> **if** the condition for Case 0 applies
>   INITIALIZATION
> **else if** the condition for Case 1 applies
>   UNREALIZATION
>   **else if** the condition for Case 2 applies
>   INSERTION INTO B_1
>   **else if** the condition for Case 3 applies
>   REALIZATION
>   **else** $\sigma \leftarrow \sigma \,\char`\^ m_\sigma$.

It's a matter of taste.[2]

We say that initialization has the highest *precedence* of the five actions, followed by unrealization, insertion, realization, and the default.

---

[1] Our **switch** statement is similar to the **switch** statement in the programming language Java.

[2] My students voted unanimously for the **switch** statement here.

14.3.1 Main Code

ALGORITHM 14.1

```
 1   B_0 ← ∅.
 2   B_1 ← ∅.
 3   for s ← 0 to ∞
 4       TP_s ← {λ}.
 5       σ ← λ.
 6       for ℓ ← 0 to s − 1
             // Node σ, which is on level ℓ of TP_s, acts. Also, the child of σ in TP_s is determined.
 7           e ← ⌊ℓ/2⌋.
 8           switch
 9               Case 0: m_σ = −1.
                         INITIALIZATION.
                         break.
10               Case 1: there is an i < m_σ(s) such that some b ≤ φ(σ, i)[s]
                         entered B_0 during a stage in [lastRealized(σ, i)[s], s).
                         UNREALIZATION.
                         break.
11               Case 2: there is an i such that inserted(σ, i) = FALSE
                         and lastRealized(σ, i)[s] < entryIntoC(i) ≤ s.
                             // Thus, n_σ(i) became C-permitted after it was last realized.
                         INSERTION INTO B_1.
                         break.
12               Case 3:    Φ_e^{B_0}(n_σ(m_σ))[s] = 0
                         and σ had no star witness at the start of stage s.
                         REALIZATION.
                         break.
13               Default:
                             σ ← σ ⌢ m_σ.

14       TP_s ← TP_s ∪ {σ}.

15       Cancel each node to the right of TP_s.

16       B_0 ← B_0 ∪ {2a_s}.     // This encodes a_s into B_0.
17       B_1 ← B_1 ∪ {2a_s}.     // This encodes a_s into B_1.
```

14.3.2 Subroutines

Here are the four subroutines (or "methods," as they are known in Java) called by the main code of Algorithm 14.1. We assume that the subroutines have both read and write access to all of the variables of the main routine.

INITIALIZATION

```
 1   n_σ(0) ← a fresh number.
 2   m_σ ← 0.
 3   σ ← σ ⌢ 0.
```

UNREALIZATION

1 Fix the least such i.
   //   We say that b *unrealizes* $n_\sigma(i)$ during stage s, where b was the first
   //   number $\leq \varphi(\sigma, i)[s]$ that entered B_0 during a stage in $[lastRealized(\sigma, i)[s], s)$.
2 **if** $n_\sigma(i) \in B_1$
3      $n_\sigma(i) \leftarrow$ a fresh number.
4 Undefine witness $n_\sigma(j)$ for each j such that $i < j \leq m_\sigma$.
5 $m_\sigma \leftarrow i$.   // $n_\sigma(i)$ is now unrealized.
6 $\sigma \leftarrow \sigma^\frown i$.

INSERTION INTO B_1

1 Fix the least such i.
2 $B_1 \leftarrow B_1 \cup \{n_\sigma(i)\}$.
3 Cancel node $\sigma^\frown i$.
4 $\sigma \leftarrow \sigma^\frown i$.

REALIZATION

1 $i \leftarrow m_\sigma$.
2 $n_\sigma(i+1) \leftarrow$ a fresh number.
3 $m_\sigma \leftarrow i + 1$.   // $n_\sigma(i)$ is now realized.
4 Cancel node $\sigma^\frown R$.   // The R branch leads to an inert node (see Note 5 below).
5 $\sigma \leftarrow \sigma^\frown R$.

14.3.3   Notes on the Algorithm

1. During the body of the **for** loop of line 6 of the main code, we visit node σ, which is on level ℓ of the tree (in other words, $|\sigma| = \ell$). We assume here that ℓ is even (that is, $\ell = 2e$); the case in which ℓ is odd (that is, $\ell = 2e + 1$) is analogous, swapping B_0 with B_1.
2. Regarding the condition for Case 0, note that

$$m_\sigma = -1 \iff n_\sigma(0) \uparrow$$
$$\iff \text{either this is the first visit to } \sigma,$$
$$\text{or this is the first visit to } \sigma \text{ since it was last canceled.}$$

3. How can we determine, in a finite amount of time, the truth of the condition for Case 2?

Note that

$$lastRealized(\sigma, i)[s] < entryIntoC(i)$$

implies $i < m_\sigma(s)$, because

$$i \geq m_\sigma(s) \implies lastRealized(\sigma, i)[s] = \infty.$$

Therefore, we need examine only finitely many values of i.
However, the function $entryIntoC(i)$ is not computable (see Exercise 8). So, for a fixed i, how can we determine the truth of

$$lastRealized(\sigma, i)[s] < entryIntoC(i) \leq s$$

in a finite amount of time?
Here's how:

```
if i ∉ Cₛ
        output(FALSE)
                // because either i ∉ C (and so lastRealized(σ, i)[s] > −1 = entryIntoC(i)),
                // or i ∈ C − Cₛ (and so entryIntoC(i) > s).
    else t ← the number such that i = cₜ. // Thus, t ≤ s  and  t = entryIntoC(i).
        if lastRealized(σ, i)[s] < t
                output(TRUE)
        else output(FALSE).
```

Because C_s is finite, this can be executed in a finite amount of time.

4. Whenever a node $\sigma \frown R$ is visited, it gets initialized, because it was just canceled by line 4 of the realization subroutine. Thus, node $\sigma \frown R$ is *inert*, that is, it can never perform an insertion. This will not hurt the overall proof, because $\sigma \frown R \notin TP$, by Fact 6. Likewise, each descendant of $\sigma \frown R$ is inert.

5. Any increase in m_σ is caused by a realization, and hence is an increase of exactly one.

6. The purpose of line 4 of the unrealization subroutine is to preserve the situation depicted in Fig. 14.3.

7. There are three wrinkles[3] in Algorithm 14.1 that probably seem mysterious at this point:

   (a) line 12 of the main code: the clause involving a star witness in the condition for Case 3 (realization).
   (b) lines 2 and 3 of the unrealization subroutine: the possible assignment of a fresh number to $n_\sigma(i)$.
   (c) lines 3–4 of the insertion subroutine: the cancellation of $\sigma \frown i$ during the insertion of $n_\sigma(i)$, followed by a visit to $\sigma \frown i$ (rather than, say, a visit to $\sigma \frown m_\sigma$).

---

[3] See the remarks about wrinkles in the Afternotes section.

We use (a) in the proof of Fact 5. Furthermore, (a) has no harmful side effects, because of an important principle employed during the design of Algorithm 14.1:

It doesn't hurt to delay realization.[4]

In other words, it's O.K. for us to slow down, *but not stop* the growth of m_σ (m_σ might stop growing on its own, but we must not force it to do so).

This principle led us to give both unrealization and insertion precedence over realization in the **switch** statement.

We use (b) in the proof of Lemma 1.

We use (c) in the proofs of Fact 3, Lemmas 1, and 2.

8. A witness $n_\sigma(i)$ becomes realized when m_σ increases to $i + 1$. One might (mistakenly) think that it becomes realized at the moment when $\Phi_e^{B_0}(n_\sigma(i))$ becomes equal to 0. Likewise, $n_\sigma(i)$ becomes unrealized when m_σ decreases to i.

9. Algorithm 14.1, and the proofs of the verification section, are so laden with details that it behooves us to step back and and look at the big picture: there's a trade-off going on here. The more cancellation done by the algorithm, the easier it is to prove that odd numbers do not cause unrealization (Fact 3). On the other hand, cancellation can only make it harder to show that certain witnesses n hang around forever and result in

$$B_1(n) = 0 \neq \Phi_e^{B_0}(n)$$

(which occurs in a case of the proof of Lemma 1).

14.4   Verification

As usual, the purpose of the verification section is to show that the constructed sets meet the requirements.

Sets B_0 and B_1 are built in stages, each of which can be executed in a finite amount of time; hence B_0 and B_1 are c.e. By coding, we obtain $A \leq_T B_0$ and $A \leq_T B_1$. Lemma 1 says that the R-requirements are met, and so $B_0 \not\leq_T B_1$ and $B_1 \not\leq_T B_0$. Lemma 2 says that $B_0 \leq_T C$ and $B_1 \leq_T C$, thereby completing the proof of (14.2) and hence of Theorem 14.

Section 14.4.3 contains facts that are useful for proving the two lemmas. Section 14.4.4 contains those lemmas.

Figure 14.4 depicts a directed graph,[5] whose vertices correspond to the results in this chapter; there is an edge from vertex v to vertex w if we use the result corresponding to v when proving the result corresponding to w.

---

[4] See Chap. 15 for more about delaying tactics.

[5] This graph is acyclic, mercifully.

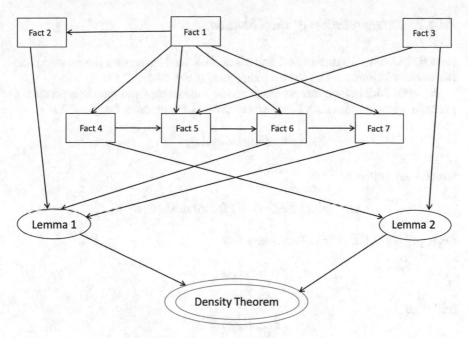

Fig. 14.4 Dependencies among the results in Chap. 14

14.4.1 More Definitions

As in all of our tree arguments, the true path is defined word-for-word as in (11.1).

We say that a node σ *is initialized* during stage s if the initialization action occurs while σ is being visited during stage s. We say that σ *performs* an unrealization (or insertion or realization) during stage s if that action occurs while σ is being visited during stage s. Likewise, we say that σ *takes the default* during stage s if the default action occurs while σ is being visited during stage s.

A witness $n_\sigma(i)$ is *ready for insertion* at a σ-stage s if

$$inserted(\sigma, i)[s] = \text{FALSE} \quad \text{and} \quad lastRealized(\sigma, i)[s] < entryIntoC(i) \leq s.$$

Even if $n_\sigma(i)$ is ready for insertion at a σ-stage s, it might not be inserted during stage s, either because σ is canceled during stage s before being visited, or because σ performs an unrealization during stage s (because unrealization takes precedence over insertion), or because $n_\sigma(j)$ is inserted during stage s for some $j < i$ (note the word "least" in line 1 of the insertion subroutine).

14.4.2 Interpretation of the Guesses

Now that we have Algorithm 14.1 and some associated terminology, we can explain the intuition behind the tree (take another look at Fig. 14.1).

As usual with tree arguments, each branch emanating from a node represents a guess. In particular, for each k, the branch labeled k represents the guess that

$$k = \liminf_s m_\sigma(s),$$

which is equivalent to

$$k = \min\{i \,:\, m_\sigma(s) = i \text{ for infinitely many } s \}$$

(as is pointed out in (9.16)). Fact 5 says that

$$\liminf_s m_\sigma(s) < \infty$$

(although

$$\max_s m_\sigma(s) = \infty$$

is possible, even if $\sigma \in TP$).

So, what guess does the righmost branch (that is, the branch labeled R) represent? The answer is "not applicable," because $\sigma \,^\frown R \in TP$ is impossible, which follows from Fact 6. The label "R" is for "realization," because a node σ takes its R branch only when it realizes a witness (see line 5 of the realization subroutine). The purpose of the R branch is to allow us to rule out realization when considering the possible actions taken at a certain node during certains stages; for example, we do this in Cases 3.3 and 4.2 of the proof of Fact 3, and in the proof of Fact 6.

As usual, the true path consists entirely of correct guesses; this follows from Fact 6.

14.4.3 Facts

Fact 14.1 *For each σ and i, there is at most one stage during which $n_\sigma(i)$ is inserted.*

Proof Suppose that $n_\sigma(i)$ is inserted during a stage s, and let $t > s$. Then

$$inserted(\sigma, i)[t] = \text{TRUE}$$

and so $n_\sigma(i)$ is not ready for insertion at stage t.

$$\textit{QED Fact 1}$$

Fact 14.2 *For each σ and i, $n_\sigma(i)$ gets a fresh number by line 3 of the unrealization subroutine at most once.*

Proof Suppose that $n_\sigma(i)$ gets a fresh number by line 3 of the unrealization subroutine during a stage s. Then $n_\sigma(i)[s] \in B_1[s]$ (assuming that $|\sigma|$ is even; if $|\sigma|$ were odd then we would substitute B_0 for B_1). Let $t > s$. If $n_\sigma(i)$ gets a fresh number by line 3 of the unrealization subroutine during stage t, then

$$n_\sigma(i)[s] \neq n_\sigma(i)[t] \in B_1[t]$$

and so $n_\sigma(i)$ is inserted both before and after stage s, contradicting Fact 1.

QED Fact 2

Fact 14.3 *Let b be a number that unrealizes a witness. Then b is even.*

Recall that the definition of a number unrealizing a witness is given in the comment after line 1 of the unrealization subroutine. Even numbers enter B_0 and B_1 by coding, whereas odd numbers enter them by the insertion subroutine; therefore, another way to phrase this result is "If b unrealizes a witness then $b/2 \in A$."

Proof Let σ, i, and u be such that b unrealizes $n_\sigma(i)$ during stage u. Assume that $|\sigma|$ is even (otherwise we would swap B_0 with B_1 in this proof); let e be such that $|\sigma| = 2e$. Thus, σ might insert some numbers into B_1, and might try to restrain some numbers from entering B_0.

Let

$$r = lastRealized(\sigma, i)[u].$$

Assume for a contradiction that b were odd. Then there exist ξ, j, and t such that

$$b = n_\xi(j)[t]$$

was inserted into B_0 during stage t. Thus,

$$b \leq \varphi(\sigma, i)[u] = \varphi_e^{B_0}(n_\sigma(i))[r] \tag{14.3}$$

and[6]

$$r \leq t < u. \tag{14.4}$$

We know $t \neq u$ because $\big[lastRealized(\sigma, i)[s], s \big)$ rather than $\big[lastRealized(\sigma, i)[s], s \big]$ is written in line 10 of the main code.

Consider five cases, based on the relative positions of nodes ξ and σ; in each case we will derive a contradiction.

---

[6] I tried to pick the names of these stages mnemonically: r is for r(ealization), t for (inser)t(ion), and u for u(nrealization). Also, I wanted their alphabetical order to match their numerical order, to help keep (14.4) in mind.

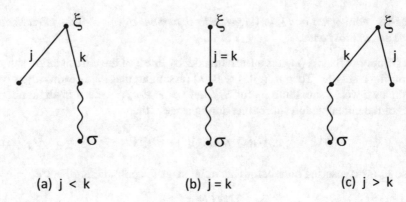

Fig. 14.5 Relative sizes of j and k, in Case 3

Case 1. $\xi <_L \sigma$.

Then σ was canceled during stage t by line 15 of the main code (because t is a ξ-stage), and so $r > t$, contradicting (14.4).

Case 2. $\sigma <_L \xi$.

Then ξ was canceled during stage r by line 15 of the main code (because r is a σ-stage). Therefore $n_\xi(j)$ was assigned the value b during a stage in (r, t) (this could not have happened during stage r, because r is not a ξ-stage). Because b was chosen to be fresh,

$$b > \varphi_e^{B_0}(n_\sigma(i))[r],$$

contradicting (14.3).

Case 3. $\xi \prec \sigma$.

It cannot be that $\xi \,\widehat{}\, R \preceq \sigma$, because otherwise σ would be canceled during every σ-stage (by line 4 of the realization subroutine), and so $n_\sigma(i)$ would never be realized. Hence, $\xi \,\widehat{}\, k \preceq \sigma$ for some k.

Note that t is a $(\xi \,\widehat{}\, j)$-stage, by line 4 of the insertion subroutine.

Case 3.1 $j < k$ (as in Fig. 14.5a).

Then σ was canceled during stage t by line 15 of the main code. Therefore $r > t$, contradicting (14.4).

Case 3.2 $j = k$ (as in Fig. 14.5b).

Then σ was canceled during stage t by line 3 of the insertion subroutine (as a reminder, whenever a node is canceled, so are all of its descendant nodes). Therefore $r > t$, contradicting (14.4).

Case 3.3 $j > k$ (as in Fig. 14.5c).

Consider the action performed by ξ during stage r (which is a ξ-stage because it is a σ-stage). It was not initialization, because otherwise σ would not have performed a realization during stage r. It was neither unrealization nor the default action, because otherwise $m_\xi(r + 1) = k < j$ and so $n_\xi(j)[r + 1]\uparrow$, and hence during a stage in (r, t), $n_\xi(j)$ would have received the fresh number

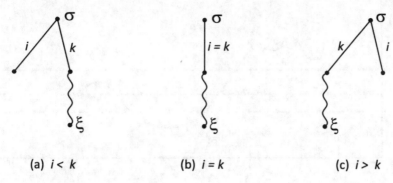

Fig. 14.6 Relative sizes of i and k, in Case 4

$$b > \varphi_e^{B_0}(n_\sigma(i))[r],$$

contradicting (14.3). Nor was it insertion, because otherwise $\xi \smallfrown k$ (and hence σ) would have been canceled (by line 3 of the insertion subroutine) during stage r, but then σ would not have performed a realization during stage r. Nor was it realization, because otherwise r would have been a $(\xi \smallfrown R)$-stage rather than a $(\xi \smallfrown k)$-stage. Thus, we have a contradiction.

So, in each of the three subcases of Case 3, we have a contradiction.

Case 4 $\sigma \prec \xi$.

It cannot be that $\sigma \smallfrown R \preceq \xi$, because otherwise ξ would be canceled during every ξ-stage (by line 4 of the realization subroutine), and so $n_\xi(j)$ would never be inserted (or even realized). Hence, $\sigma \smallfrown k \preceq \xi$ for some k.

Case 4.1 $i < k$ (as in Fig. 14.6a).

By (14.3), $n_\xi(j)$ was assigned the fresh value b during some stage $r_0 < r$ ($r_0 = r$ is impossible because r is a $\sigma \smallfrown R$-stage and hence is not a ξ-stage). Therefore r_0 is a ξ-stage and hence

$$m_\sigma(r_0) \geq k.$$

Furthermore,

$$m_\sigma(r) = i < k,$$

because $n_\sigma(i)$ became realized during stage r. Let r_1 be the largest number in $[r_0, r)$ such that

$$m_\sigma(r_1) \geq k \quad \text{and} \quad m_\sigma(r_1 + 1) < k. \tag{14.5}$$

See Fig. 14.7, in which $k_1 < i$ (although $k_1 \geq i$ is possible, too). The existence of such an r_1 follows from what could be regarded as a discrete version of the mean-value theorem. The variable m_σ decreased to some $k_1 < k$ during stage r_1. Therefore during stage r_1, either node σ was initialized (in which

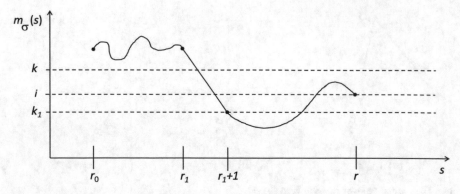

Fig. 14.7 Stages r_0 and r_1 in Case 4.1

case $k_1 = 0$), or witness $n_\sigma(k_1)$ became unrealized. Hence $\sigma {}^\frown k_1 \in TP_{r_1}$ and so node $\sigma {}^\frown k$ (and therefore ξ) was canceled during stage r_1 by line 15 of the main code. After stage r_1, node ξ was not visited until after stage r, because r_1 is defined as the *largest* number in $[r_0, \ r)$ satisfying (14.5); this contradicts (14.3).

Case 4.2 $i \geq k$ (as in Fig. 14.6b or c).

Because t is a ξ-stage, it is also a $(\sigma {}^\frown k)$-stage. Consider the action performed by σ during stage t. It was neither initialization nor the unrealization of $n_\sigma(k)$ nor the default action, because otherwise

$$m_\sigma(t + 1) = k \leq i$$

and so $n_\sigma(i)$ would have been either unrealized or undefined at the end of stage $t + 1$, hence we would have $r > t$, contradicting (14.4). Nor was it the insertion of $n_\sigma(k)$, because that would have canceled ξ by line 3 of the insertion subroutine, and so ξ could not have performed an insertion during stage t. Nor was it realization, because otherwise t would have been a $(\sigma {}^\frown R)$-stage rather than a $(\sigma {}^\frown k)$-stage. Thus, none of the five actions could have been performed by σ during stage t.

So, in both of the subcases of Case 4, we have a contradiction.

Case 5. $\sigma = \xi$.

This is impossible, because ξ is on an odd level of the tree (because it inserted a number into B_0), whereas σ is on an even level.

Thus, in each of the five cases we have a contradiction. Therefore, b is even.

QED Fact 3

Fact 14.4 *Suppose that σ is visited infinitely often, that $b = n_\sigma(k)[s]$ is ready for insertion at a σ-stage s, and that*

$$(\forall z \geq s)[\, k < m_\sigma(z)\,]. \tag{14.6}$$

Then during some stage $t \geq s$, b is inserted (into B_1 if $|\sigma|$ is even; otherwise into B_0) and is a permanent star witness at stage $t + 1$.

Proof It must be that

$$inserted(\sigma, i)[s] = \text{FALSE}$$

because $n_\sigma(k)$ is ready for insertion at stage s. Assume for a contradiction that $n_\sigma(k)$ is never inserted during or after stage s. Then, during each subsequent σ-stage, witness $n_\sigma(k)$ remains ready for insertion, because (by (14.6))

$$\big(\forall t > s\big)\big[\, lastRealized(\sigma, k)[t] = lastRealized(\sigma, k)[s]\,\big].$$

So, what prevents $n_\sigma(k)$ from being inserted during a σ-stage greater than or equal to s? Only three actions of σ take precedence over the insertion of $n_\sigma(k)$:

(a) the initialization of σ,
(b) the unrealization of a witness of σ,
(c) the insertion of $n_\sigma(i)$ for some $i < k$.

Action (a) does not occur during or after stage s, by (14.6). Action (c) occurs at most k times, by Fact 1. Hence, there is a stage $x \geq s$ during or after which action (b) occurs during each σ-stage. However, σ cannot perform an infinite sequence of unrealizations while performing no realizations (over which insertion takes precedence), because m_σ cannot decrease infinitely often while never increasing. Hence, because there are infinitely many σ-stages, there exists a σ-stage $y \geq x$ during which neither (a) nor (b) nor (c) occurs. Witness $n_\sigma(k)$ is inserted during stage y. This is a contradiction.

Thus, $b = n_\sigma(k)[s]$ is inserted during a stage $t \geq s$. No number $b_0 \leq \varphi_e^{B_0}(k)[t]$ is inserted into B_0 during or after stage t, by (14.6). In other words,

$$B_0[t] \upharpoonright \varphi_e^{B_0}(k)[t] = B_0 \upharpoonright \varphi_e^{B_0}(k)[t]$$

and so b is a permanent star witness, starting at stage $t + 1$.

QED Fact 4

It would be convenient if, for each $\sigma \in TP$, the function m_σ had a finite maximum. Unfortunately, that might not be true. However, Fact 5 says that m_σ does have a finite liminf, and that will suffice for our purposes. Actually, Fact 5 is a tad stronger than we need, because it holds for each σ, regardless of whether σ is on the true path.[7]

---

[7] We don't benefit from this extra strength, because Fact 5 is used only in the proof of Fact 6, which concerns only nodes in *TP*.

Fact 14.5 *For each σ,*

$$\liminf_s m_\sigma(s) < \infty.$$

Proof Fix σ. Assume for a contradiction that

$$\liminf_s m_\sigma(s) = \infty. \tag{14.7}$$

Then σ is visited infinitely often, because m_σ can increase only when σ is visited. We will show $C \leq_T A$ by describing an algorithm that, using an oracle for A, decides membership in C.

Let x be such that

$$(\forall s \geq x)\big[\, m_\sigma(s) \geq 1 \,\big]$$

(such an x exists by (14.7)). Node σ is never initialized during or after stage x.

The following algorithm (which knows σ and x) is given p as input, and determines whether $p \in C$; it uses an oracle for A.

> ALGORITHM 14.2
> 1 $y \leftarrow$ some number such that
> 2 $y > x$,
> 3 $p < m_\sigma(y)$,
>   and
> 4 $\neg\big(\exists s \geq y\big)\big[\, 2a_s \leq \max\big\{\varphi(\sigma, i)[y] : i \leq p\big\} \,\big].$
> 5 **if** $p \in C_y$
> 6   output("p is in C")
> 7 **else** output("p is not in C").

In line 4, we multiply a_s by 2 because members of A are doubled when inserted by lines 16 and 17 of the main code of Algorithm 14.1. Informally, line 4 says that during or after stage y, no *even* number gets inserted and subsequently unrealizes $n_\sigma(i)[y]$ for some $i \leq p$.

The claim $C \leq_T A$ follows from the following propositions.

Proposition 1 *Using an oracle for A we can determine, in a finite amount of time, whether a given y satisfies lines 2 through 4 of Algorithm 14.2.*

Proof It is trivial to determine the truth of lines 2 and 3. To determine the truth of line 4, we do the following:

$a \leftarrow 0.$
answer $\leftarrow$ TRUE.
while *answer* = TRUE and $2a \leq \max \{ \varphi(\sigma, i)[y] : i \leq p \}$
   if $a \in A - A_{y-1}$ // Here is where we use the oracle for A.
      answer $\leftarrow$ FALSE.
   $a \leftarrow a + 1.$
output(*answer*).

The **while** loop iterates no more than

$$1 + \max \{ \varphi(\sigma, i)[y] : i \leq p \}/2$$

times.   □

Proposition 2 *There exists y that satisfies lines 2 through 4 of Algorithm 14.2.*

Proof By (14.7), there exists $y > x$ such that

$$(\forall s \geq y)\big[\, p < m_\sigma(s) \,\big]. \tag{14.8}$$

Thus, y satisfies lines 2 and 3 of Algorithm 14.2.
   We claim that y satisfies line 4 of Algorithm 14.2; assume the contrary. Then

$$\big(\exists s \geq y\big)\big[\, 2a_s \leq \max \{ \varphi(\sigma, i)[y] : i \leq p \} \,\big],$$

which is equivalent to

$$\big(\exists i \leq p\big)\big(\exists s \geq y\big)\big[\, 2a_s \leq \varphi(\sigma, i)[y] \,\big];$$

let i' be the least such i, and let $z \geq y$ be such that

$$2a_z \leq \varphi(\sigma, i')[y]. \tag{14.9}$$

Then

$$\varphi(\sigma, i')[z] = \varphi(\sigma, i')[y], \tag{14.10}$$

because otherwise

$$lastRealized(\sigma, i')[z] \neq lastRealized(\sigma, i')[y],$$

which would imply that $n_\sigma(i')$ became unrealized during a stage in $[y, z)$, which would contradict (14.8) because $i' \leq p$.
   During stage z, the number $2a_z$ is inserted by lines 16 and 17 of Algorithm 14.1. By (14.9) and (14.10),

$$2a_z \leq \varphi(\sigma, i')[z].$$

Therefore, during the first σ-stage $z' > z$, the condition for unrealization (in Case 1 of the **switch** statement) is satisfied. Hence, during stage z', witness $n_\sigma(i')$ becomes unrealized, because

(i) σ is never initialized after stage x,
(ii) unrealization takes precedence over every other action except for initialization,
(iii) i' was defined as the *least* such i.

Therefore

$$m_\sigma(z' + 1) = i' \le p,$$

contradicting (14.8).

Thus, y satisfies line 4 of Algorithm 14.2.   □

Let y be the number computed by Algorithm 14.2 (it exists by Proposition 2).

Proposition 3 *The output of Algorithm 14.2 is correct, that is,*

$$p \in C \iff p \in C_y.$$

Proof Assume the contrary; then $p \in C - C_y$. In other words,

$$y < entryIntoC(p).\tag{14.11}$$

Let $z \ge y$. During stage z:

(i) σ is not initialized (by line 2 of Algorithm 14.2, because σ is never initialized after stage x),
(ii) for each $i \le p$, no even number unrealizes $n_\sigma(i)$ (by line 4 of Algorithm 14.2),
(iii) for each $i \le p$, no odd number unrealizes $n_\sigma(i)$ (by Fact 3).

Therefore, by line 3 of Algorithm 14.2,

$$(\forall z \ge y)\big[\, p < m_\sigma(z) \,\big].\tag{14.12}$$

Witness $n_\sigma(p)$ does not receive C-permission to be inserted before stage $entryIntoC(p)$. Let s be the first σ-stage such that $s \ge entryIntoC(p)$ (such an s exists because σ is visited infinitely often, as is pointed out near the start of the proof of Fact 5). Then

$$inserted(\sigma, \, p)[s] = \text{FALSE}.$$

By (14.11),

$$y < entryIntoC(p) \le s.$$

Hence

$$lastRealized(\sigma, p)[s] < y \qquad \text{(by (14.12), because } s \geq y)$$
$$< entryIntoC(p)$$
$$\leq s.$$

Thus, $n_\sigma(p)$ is ready for insertion at stage s. Therefore, by (14.12) and Fact 4, $n_\sigma(p)$ eventually becomes a permanent star witness. Hence σ performs only finitely many realizations, by the condition for realization in Case 3 of the **switch** statement, and so

$$\liminf_s m_\sigma(s) < \infty,$$

contradicting (14.7).

Therefore the output of Algorithm 14.2 is correct.   □

Propositions 1, 2, and 3 together imply that Algorithm 14.2 correctly determines whether $p \in C$, in a finite amount of time, using an oracle for A. Therefore $C \leq_T A$, contradicting the assumption $A <_T C$ of Theorem 14. Thus, the assumption (14.7) has led to a contradiction.

QED Fact 5

Fact 14.6 *Let* $\sigma \in TP$, *and let*

$$k = \liminf_s m_\sigma(s) \tag{14.13}$$

(the existence of this liminf follows from Fact 5). Then $\sigma {\,}^\frown k \in TP$.

Proof If σ is initialized infinitely often[8] then $k = 0$ and $\sigma {\,}^\frown k \in TP$, and so this proof would be finished. So assume that σ is initialized only finitely often. Let x be such that

  (i) σ is not initialized during or after stage x,
 (ii) $(\forall z \geq x)\big[\ m_\sigma(z) \geq k\ \big]$,
(iii) $(\forall i \leq k)\big[\ n_\sigma(i)$ is not inserted during or after stage $x\ \big]$.

Such an x exists by

  (i) our assumption that σ is initialized only finitely often,
 (ii) Equation (14.13),
(iii) Fact 1, applied to each $i \leq k$.

Let $i < k$. Assume for a contradiction that some $y > x$ is a $(\sigma {\,}^\frown i)$-stage. Consider the action performed by σ during stage y. It was not initialization, by condition (i)

---

[8] This is impossible, by Fact 7. However, we cannot use Fact 7 here, because our proof of Fact 7 uses Fact 6.

on the choice of x. Nor was it unrealization, because otherwise $m_\sigma(y+1) = i < k$, contradicting condition (ii) on the choice of x. Nor was it insertion, by condition (iii) on the choice of x. Nor was it realization, because otherwise y would be a $(\sigma ^\frown R)$-stage. Nor was it the default action, because otherwise $m_\sigma(y+1) = i < k$, contradicting condition (ii) on the choice of x. Therefore, no $(\sigma ^\frown i)$-stage is greater than x, and so there are only finitely many $(\sigma ^\frown i)$-stages. Hence, because $\sigma \in TP$,

$$(\forall \rho \lhd \sigma ^\frown k)[\rho \text{ is visited only finitely often}].$$

To prove that $\sigma ^\frown k \in TP$, it remains only to show that there are infinitely many $(\sigma ^\frown k)$-stages. Assume for a contradiction that there is a stage t larger than each $(\sigma ^\frown k)$-stage. Let

$$z > \max\{x, t\}$$

be a σ-stage such that

$$m_\sigma(z) = k.$$

Such a z exists because, by (14.13), there are infinitely many values of s such that $m_\sigma(s) = k$, and the variable m_σ can change only when σ is visited or canceled. Consider the action performed by σ during stage z. It was not initialization, by condition (i) on the choice of x. Nor was it unrealization, because otherwise $m_\sigma(z + 1) < k$, contradicting condition (ii) on the choice of x. Nor was it insertion, by condition (iii) on the choice of x. Nor was it the default action, because otherwise z would be a $(\sigma ^\frown k)$-stage, contradicting $z > t$ and the choice of t.

Nor was it realization; to see this, assume the contrary. Then $m_\sigma(z + 1) = k + 1$. By (14.13), there is a least stage $z' > z$ during which σ performs an unrealization on the witness $n_\sigma(k)$. Hence z' is a $(\sigma ^\frown k)$-stage, contradicting $z' > t$ and the choice of t.

Thus, the assumption that there are only finitely many $(\sigma ^\frown k)$-stages has led to a contradiction. Therefore

$$\sigma ^\frown k \in TP.$$

QED Fact 6

Fact 6 has a number of nice corollaries, such as:

1. For each σ, $\sigma ^\frown R \notin TP$.
2. The set TP is an infinite path. This can be proved by ordinary induction on the level of the tree, as follows. The root of the tree is in TP. By Fact 6, if $\sigma \in TP$ then σ has a child in TP.
3. Along the true path all of the guesses (as described in Sect. 14.4.2) are correct.

Fact 14.7 *Let $\sigma \in TP$. Then σ is canceled only finitely often.*

Proof Cancellation occurs in three places: line 3 of the insertion subroutine, line 4 of the realization subroutine, and line 15 of the main code.

Line 3 of the insertion subroutine cancels σ when witness $n_\tau(i)$ is inserted for some τ and i such that $\tau {}^\frown i \preceq \sigma$. This occurs only finitely often, by Fact 1, because σ has only finitely many ancestors.

Line 4 of the realization subroutine never cancels σ, because otherwise $\tau {}^\frown R \preceq \sigma$ for some τ, and so $\tau {}^\frown R \in TP$, contradicting Fact 6.

We now show by induction on $|\sigma|$ that line 15 of the main code cancels σ only finitely often. In other words, we show that the set

$$L_\sigma = \{s : TP_s <_L \sigma\}$$

is finite.

As the basis for the induction, note that

$$L_\Lambda = \emptyset.$$

For the inductive step, suppose that τ is the parent of σ, and assume the claim for τ; that is, assume that L_τ is finite. Let $s \in L_\sigma$; then $TP_s <_L \tau$ is impossible. Therefore, either $TP_s <_L \tau$ or $\tau \in TP_s$.

Case 1. $TP_s <_L \tau$ (as in Fig. 14.8a).
There are only finitely many such s, because L_τ is finite.
Case 2. $\tau \in TP_s$ (as in Fig. 14.8b).
By Fact 6 applied to τ, we have $\sigma = \tau {}^\frown k$, where $k = \liminf_z m_\tau(z)$.
Assume for a contradiction that there were infinitely many such s. Then at least one of the nodes in the finite set

$$\{\tau {}^\frown 0, \ \tau {}^\frown 1, \ \ldots, \ \tau {}^\frown (k-1)\}$$

is visited infinitely often, contradicting $\tau {}^\frown k \in TP$.

QED Fact 7

Fig. 14.8 Cases in the proof of Fact 7

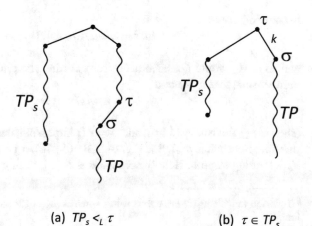

(a) $TP_s <_L \tau$ (b) $\tau \in TP_s$

14.4.4 Lemmas

Lemma 14.1 *Each R-requirement is met.*

Proof Fix e. We will show that R_{2e} is met, the proof for R_{2e+1} being analogous. By Fact 6, *TP* is an infinite path, so there is a node σ on level $2e$ of *TP*. Also by Fact 6, $\sigma \hat{\ } k \in TP$ where

$$k = \liminf_s m_\sigma(s). \tag{14.14}$$

When is $n_\sigma(k)$ assigned a fresh value? It occurs when

(a) σ is initialized (line 1 of the initialization subroutine), if $k = 0$.
(b) $n_\sigma(k)$ becomes unrealized when $n_\sigma(k) \in B_1$ (line 3 of the unrealization subroutine).
(c) $n_\sigma(k - 1)$ becomes realized (line 2 of the realization subroutine).

By Fact 7, (a) occurs only finitely often. By Fact 2, (b) occurs at most once. If (c) occurs during a stage s then $m_\sigma(s) = k - 1$; therefore by (14.14), (c) occurs only finitely often. To summarize, the value of $n_\sigma(k)$ changes only finitely often.

Therefore $\lim_s n_\sigma(k)[s]$ exists[9]; call it n. Let x be the stage during which variable $n_\sigma(k)$ is assigned the value n. Thus

$$(\forall s > x)\big[\, n = n_\sigma(k)[s] \,\big]. \tag{14.15}$$

Therefore

$$(\forall s > x)\big[\, k \leq m_\sigma(s) \,\big], \tag{14.16}$$

because $m_\sigma(s) < k$ would imply $n_\sigma(k)[s] \uparrow$.

We claim that $n \notin B_1$; assume the contrary. Then there was a stage t during which $n = n_\sigma(k)[t]$ was inserted into B_1. Note that $t > x$. Furthermore,

$$k < m_\sigma(t),$$

because otherwise

$$lastRealized(\sigma, k)[t] = \infty$$

and so $n_\sigma(k)$ would not have been ready for insertion at stage t. Let v be the least number such that $v > t$ and

$$k = m_\sigma(v + 1)$$

(such a $v \geq t$ exists by (14.4), and $v = t$ is impossible because the insertion subroutine does not change m_σ). See Fig. 14.9. By (14.16), σ performed an unrealization on $n_\sigma(k)$ during stage v. Hence, because $n \in B_1[v]$, during stage v the value of $n_\sigma(k)$

---

[9] Thus, the variable $n_\sigma(k)$ has a final value, whereas m_σ might not (although it does have a finite liminf).

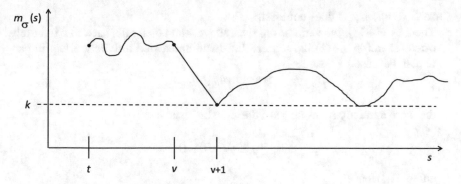

Fig. 14.9 The stages t and v in the proof of Lemma 1

was changed from n by line 3 of the unrealization subroutine, contradicting (14.15).[10] Thus, $n \notin B_1$; in other words,

$$B_1(n) = 0. \tag{14.17}$$

Case 1. $\lim_s m_\sigma(s)$ exists.

Then $k = \lim_s m_\sigma(s)$, by (14.14). Let $y > x$ be such that

$$(\forall s \geq y)\big[\, k = m_\sigma(s) \,\big].$$

Case 1.1 σ has a star witness at the start of stage y.

That star witness is $n_\sigma(i)[y]$ for some $i < k$. It is permanent, because m_σ never changes during or after stage y. Hence R_{2e} is met.

Case 1.2 σ has no star witness at the start of stage y.

If $\Phi_e^{B_0}(n)[y] = 0$ then σ would perform a realization during the first σ-stage greater than or equal to y (look again at the condition for realization in Case 3 of the **switch** statement), which would increment m_σ, a contradiction. Therefore

$$\Phi_e^{B_0}(n)[z] \neq 0$$

for each of the infinitely many σ-stages greater than or equal to y. Hence

$$\Phi_e^{B_0}(n) \neq 0, \tag{14.18}$$

by the contrapositive of the Permanence Lemma.

By (14.17) and (14.18),

$$B_1(n) = 0 \neq \Phi_e^{B_0}(n)$$

and so R_{2e} is met.

---

[10] The sole purpose of lines 2 and 3 of the unrealization subroutine is to ensure that $n \notin B_1$ here.

Case 2. $\lim_s m_\sigma(s)$ does not exist.

Then, by (14.14), the variable m_σ rises above k and then falls back to k infinitely often. Therefore $n_\sigma(k)$ becomes realized and unrealized infinitely often. In particular, because

$$n = \lim_s n_\sigma(k)[s],$$

there are infinitely many stages z during which some

$$b_0 \leq \varphi_e^{B_0}(n)[z]$$

enters B_0. Hence

$$\Phi_e^{B_0}(n) \uparrow,$$

by the contrapositive of the Permanence Lemma. Therefore

$$B_1(n) \neq \Phi_e^{B_0}(n)$$

and so R_{2e} is met.

Thus, in either Case 1 or Case 2, R_{2e} is met.

QED Lemma 1

Lemma 14.2 $B_0 \leq_T C$ and $B_1 \leq_T C$.

Proof   We will show $B_1 \leq_T C$, the proof of $B_0 \leq C$ being analogous.

Algorithm 14.3 is given an input b_1, and uses an oracle for C to determine whether $b_1 \in B_1$. The oracle for C also functions as an oracle for A, because $A <_T C$.

It calls three subroutines: The subroutine *ReportYes* outputs "$b_1 \in B_1$" and then halts the main code. The subroutine *ReportNo* outputs "$b_1 \notin B_1$" and likewise halts the main code.

The third subroutine, *cannotVisitNodeξ*(s_2), called in line 23, is more complicated. It returns a boolean value. In particular, it returns TRUE if and only if there exist σ, i, and i' such that

$$i < i' \quad \text{and} \quad \sigma{^\frown}i \preceq \xi \quad \text{and} \quad \sigma{^\frown}i' \in TP_{s_2} \qquad \text{(see Fig. 14.10)}$$

and

$$\neg(\exists s \geq s_2)\big[\, 2a_s \leq \varphi(\sigma, i)[s_2] \,\big]. \tag{14.19}$$

In (14.19), we multiply a_s by 2 because members of A are doubled when inserted by lines 16 and 17 of the main code of Algorithm 14.1.[11] Informally, (14.19) says that during or after stage s_2, no *even* number gets inserted and subsequently unrealizes $n_\sigma(i)[s_2]$ (no odd number does, either, by Fact 3).

---

[11] Equation (14.19) is similar to line 4 of Algorithm 14.2, which was used in the proof of Fact 5.

ALGORITHM 14.3

1   **if** b_1 is even
2   **if** $b_1/2 \in A$ // Use the oracle for A to determine this.
3   *ReportYes*   // Line 17 of the main code of Algorithm 14.1 puts b_1 into B_1.
4   **else** *ReportNo*.

5   Run Algorithm 14.1 until it chooses a witness $b_2 \geq b_1$.
6   **if** $b_2 > b_1$
7   *ReportNo*.   // See Note 1 below.

8   $\xi, j, s_1 \leftarrow$ values such that $n_\xi(j)$ is assigned the value b_1 during stage s_1 of Algorithm 14.1.
9   **if** $|\xi|$ is odd
10   *ReportNo*.   // Node ξ might insert b_1 into B_0,
                       // but it cannot insert anything into B_1.

11 **if** $inserted(\xi, j)[s_1] = \text{TRUE}$ // Witness $n_\xi(j)$ will never be ready for insertion
                                                     // during or after stage s_1. Therefore $b_1 \notin B_1$.
12   *ReportNo*.

13 **if** $j \notin C$ // Use the oracle for C to determine this.
14   *ReportNo*.   // b_1 will never be C-permitted.
15 Compute $entryIntoC(j)$ by looking through $c_0, c_1, \ldots$ until finding
    the value of s such that $j = c_s$, and then $entryIntoC(j) = s$.

    // See Note 2, regarding the window of opportunity for inserting b_1.
16 **if** $entryIntoC(j) < s_1$ or $n_\xi(j)[entryIntoC(j)]$ is either unrealized or unequal to b_1
17   *ReportNo*.   // See Note 3.

    // The window of opportunity for inserting b_1 is now open.
18 **for** $s_2 \leftarrow entryInto(j)$ **to** ∞
19   **if** $b_1 \in B_1[s_2]$
20   *ReportYes*. // $b_1 \in B_1[s_2] \subseteq B_1$.
21   **if** $m_\xi(s_2) \leq j$ // $n_\xi(j)$ is now either unrealized or undefined.
22   *ReportNo*. // The window of opportunity for inserting b_1 has closed.
23   **if** $cannotVisitNode\xi(s_2)$
24   *ReportNo*. // See Note 4 and Exercise 16.

Notes on Algorithm 14.3:

1. Because witnesses are always fresh when chosen, they are chosen in increasing order; hence if $b_2 > b_1$ (line 7) then b_1 is never chosen as a witness, and so we know that $b_1 \notin B_1$. Otherwise ($b_2 = b_1$), we continue.
2. Informally, strait is the gate through which b_1 must pass before entering B_1.[12] The variable $n_\xi(j)$ must remain equal to b_1 until j enters C during a stage when $j < m_\xi$ (in other words, when $n_\xi(j)$ is realized). This opens a "window of opportunity" wherein b_1 is ready for insertion into B_1 during the next ξ-stage (see Fig. 14.11). Even during that stage and during subsequent ξ-stages, b_1 might not be inserted, because that action might have to defer to an action of higher precedence. So b_1 waits. However, as it waits, if $n_\xi(j)$ becomes undefined or even unrealized (in

---

[12] Nevertheless, infinitely many odd numbers do find their way into B_1 (see Exercise 5).

other words, m_ξ becomes less than or equal to j), then the window of opportunity for inserting b_1 closes, never to reopen, because forever after that

$$lastRealized(\xi, j) \geq entryIntoC(j) \qquad (14.20)$$

and so $n_\xi(j)$ is never again ready for insertion.
3. Consider the condition in line 16.
   If $entryIntoC(j) < s_1$ then b_1 is never C-permitted, because $n_\xi(j)$ was assigned the value b_1 during stage s_1, and so forever after that (14.20) is true. Therefore, the window of opportunity for inserting b_1 never opens.
   If $n_\xi(j)[entryIntoC(j)]$ is unrealized then forever after that (14.20) is true. Therefore, the window of opportunity for inserting b_1 never opens.
   If $n_\xi(j)[entryIntoC(j)] \neq b_1$ then then b_1 will never again be a witness for any node during or after stage $entryIntoC(j)$, because witnesses are fresh when chosen.
4. Why is the boolean function $cannotVisitNode\xi(s_2)$ so-named? In the proof of Proposition 5, we will argue that if line 24 is executed then (because $cannotVisitNode\xi(s_2) = \text{TRUE}$), node ξ cannot be visited during or after stage s_2, which implies that $b_1 \notin B_1$. This is just an overview of that argument.

Proposition 4 implies that Algorithm 14.3 can be implemented to run in a finite amount of time. Proposition 5 implies the correctness of the output of Algorithm 14.3, thereby completing the proof of Lemma 2, and hence of Theorem 14.

Proposition 4 *The **for** loop in line 18 of Algorithm 14.3 does not run forever.*

Proof Assume the contrary; that is, assume that line 18 of Algorithm 14.3 is reached, but lines 20, 22, and 24 are never executed.

Consider three cases, which depend on the position of ξ relative to the true path:

Case 1. $TP <_L \xi$.
Then ξ is canceled infinitely often, by line 15 of the main code of Algorithm 14.1. Let $s \geq entryIntoC(j)$ be a stage during which such a cancellation occurs. During stage s, witness $n_\xi(j)$ is set to undefined, and so

$$j < m_\xi(s + 1).$$

Therefore, on iteration $s_2 = s + 1$ of the **for** loop of Algorithm 14.3, line 22 is executed, a contradiction.

Case 2. $\xi \in TP$.
Then there are infinitely many ξ-stages.
Line 17 did not execute (because otherwise line 18 would not have been reached), so

$$n_\xi(j)[entryIntoC(j)] = b_1. \qquad (14.21)$$

Fig. 14.10 Part of the condition for *cannotVisitNodeξ(s_2)*

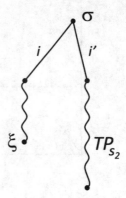

Line 22 did not execute, so

$$(\forall s \geq entryIntoC(j))[m_\xi(j)[s] > j].\tag{14.22}$$

Together, (14.21) and (14.22) imply

$$(\forall s \geq entryIntoC(j))[n_\xi(j)[s] = b_1].$$

Therefore, because line 20 is never executed,

$$(\forall s \geq entryIntoC(j))[inserted(\xi, j)[s] = \text{FALSE}].$$

Hence, for each $s \geq entryIntoC(j)$,

$$lastRealized(\xi, j)[s] < entryIntoC(j) \leq s.$$

Therefore, $b_1 = n_\xi(j)$ is ready for insertion at each ξ-stage that is greater than or equal to $entryIntoC(j)$ (informally, the window of opportunity for inserting b_1 never closes).

Hence, (14.22) and Fact 4 together imply that $b_1 = n_\xi(j)$ is eventually inserted into B_1 during some stage t, and so on iteration $s_2 = t + 1$ of the **for** loop of Algorithm 14.3, line 20 is executed, a contradiction.

Case 3. $\xi <_L TP$.

Then there exists a node σ, with children β and γ, such that $\beta \lhd \gamma$ and $\beta \preceq \xi$ and $\gamma \in TP$. Informally, the path from the root to ξ diverges from TP at node σ. Note that $\beta \neq \sigma \frown R$, because $\sigma \frown R$ is the rightmost child of σ. Also note that $\gamma \neq \sigma \frown R$, by Fact 6, because $\gamma \in TP$. Thus, there exist i and i' such that

$$i < i' \quad \text{and} \quad \beta = \sigma \frown i \quad \text{and} \quad \gamma = \sigma \frown i'$$

(see Fig. 14.12, and compare it to Fig. 14.10).

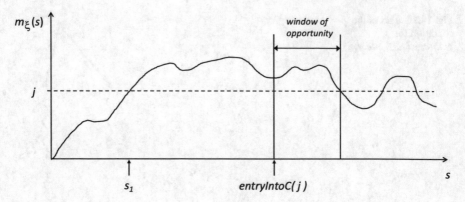

Fig. 14.11 The window of opportunity for inserting b_1

Fig. 14.12 $\xi <_L TP$

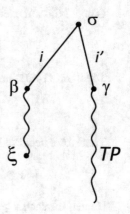

Let $t > entryIntoC(j)$ be a γ-stage during or after which no node to the left of γ is visited (such a t exists because $\gamma \in TP$). Then $\varphi(\sigma, i)$ never changes during or after stage t. Let

$$u = \lceil \varphi(\sigma, i)[t]/2 \rceil,$$

and let $s_2 > t$ be such that

$$A_{s_2-1} \restriction u = A \restriction u. \tag{14.23}$$

Note that $\varphi(\sigma, i)[s_2] = \varphi(\sigma, i)[t]$ and so

$$u = \lceil \varphi(\sigma, i)[s_2]/2 \rceil.$$

Therefore, by (14.23),

$$\neg (\exists s \geq s_2)\big[\, a_s \leq \lceil \varphi(\sigma, i)[s_2]/2 \rceil \,\big],$$

which implies (14.19). Therefore, on iteration s_2 of the **for** loop of Algorithm 14.3, line 24 is executed, a contradiction.

Thus, in each of the three cases, we have a contradiction.

QED Proposition 4

Proposition 5 *The output of Algorithm 14.3 is correct.*

Proof In the comments within Algorithm 14.3, and in the Notes on that algorithm, we have already justified each call to *Report Yes* or to *Report No* up through line 22. It remains only to justify the call to *Report No* in line 24.

Assume that line 24 executes during a stage s_2. Then

$$cannotVisitNode\xi(s_2) = \text{TRUE}.$$

Therefore we have the situation depicted in Fig. 14.10, and also (14.19) is true.

We need to prove that $b_1 \notin B_1$.

Assume for a contradiction that

$$b_1 \in B_1.$$

Then there is a stage s_4 during which b_1 is inserted into B_1. Note that s_4 is a ξ-stage and hence a $(\sigma \frown i)$-stage. Because line 20 did not execute, we have

$$b_1 \notin B_1[s_2]$$

and so

$$s_2 \le s_4.$$

Furthermore, $s_2 \neq s_4$ (because s_2 is not a ξ-stage; again look at Fig. 14.10). Therefore

$$s_2 < s_4.$$

Node ξ was not canceled during a stage in $[s_1, s_4]$, because $b_1 = n_\xi(j)$ is inserted into B_1 during stage s_4.

Which action did σ perform during stage s_4? It was not insertion, because that would have canceled ξ. Nor was it realization, because otherwise $\sigma \frown R \in TP_{s_4}$ (which is impossible, because s_4 is a ξ-stage). Nor was σ initialized during stage s_4, because if it were then b_1 would not equal $n_\xi(j)$ when ξ is visited during stage s_4, which would imply $b_1 \notin B_1$.

Therefore σ either took the default action or performed an unrealization on $n_\sigma(i)$ during stage s_4; either way,

$$m_\sigma(s_4 + 1) = i. \tag{14.24}$$

See Fig. 14.13 (in which σ took the default action during stage s_4).

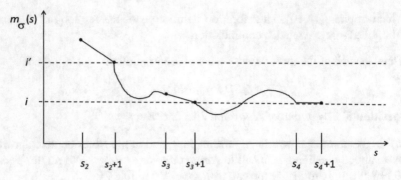

Fig. 14.13 The stages s_2, s_3, and s_4 in the proof of Case 3 of Proposition 5

Node σ did not perform a realization during stage s_2, because $\sigma \frown R \notin TP_{s_2}$. Furthermore, realization is the only action that would increase m_σ. Therefore

$$m_\sigma(s_2) \geq i' > i.$$

Hence there is a *least* stage $s_3 \geq s_2$ such that

$$m_\sigma(s_3) > i \quad \text{and} \quad m_\sigma(s_3 + 1) \leq i.$$

We have $s_3 \leq s_4$, by (14.24). Thus

$$s_1 < s_2 \leq s_3 \leq s_4.$$

Which action did σ perform during stage s_3?

It must have been either initialization or unrealization, because $m_\sigma(s_3)$ decreased during stage s_3. It was not initialization, because otherwise σ (and hence ξ) would have been canceled during a stage in $[s_2, s_3]$.

Therefore σ performed an unrealization during stage s_3. In particular, it must have been the unrealization of $n_\sigma(i)$, because if it were the unrealization of $n_\sigma(i_0)$ for some $i_0 < i$ then ξ would have been canceled during stage s_3 by line 15 of the main code of Algorithm 14.1.

Let b be the number that unrealized $n_\sigma(i)$ during stage s_3; then

$$b \leq \varphi(\sigma, i)[s_3]. \tag{14.25}$$

Because $n_\sigma(i)$ stayed realized from the start of stage s_2 to the start of stage s_3, we have

$$lastRealized(\sigma, i)[s_3] < s_2.$$

Hence

$$lastRealized(\sigma, i)[s_3] = lastRealized(\sigma, i)[s_2].$$

Therefore

$$\varphi(\sigma, i)[s_3] = \varphi(\sigma, i)[s_2],$$

and so, by (14.25),

$$b \leq \varphi(\sigma, i)[s_2].$$

By Fact 3, b is even. Hence, there exists $s \geq s_2$ such that

$$b = 2a_s \leq \varphi(\sigma, i)[s_2]$$

(see Exercise 17), contradicting (14.19). Thus, our assumption that $b_1 \in B_1$ has led to a contradiction.

QED Proposition 5

By Proposition 5, $B_1 \leq_T C$.

QED Lemma 2

The theorem follows from Lemmas 1 and 2.

QED Theorem 14

14.5 What's New in This Chapter?

1. Coding. It's a simple way to guarantee that $A \leq_T B$ where A is a given c.e. set and B is a c.e. set under construction. In this chapter, we encoded the members of A as the even numbers of B_0 and B_1, thereby leaving us free to use the odd numbers of B_0 and B_1 to meet other requirements.

2. A way to guarantee that $B \not\leq_T A$, where A is a given c.e. set and B is a c.e. set under construction. In particular, we used the Friedberg-Muchnik method to build incomparable c.e. sets B_0 and B_1, while using coding to guarantee that $A \leq_T B_0$ and $A \leq_T B_1$. Therefore, we had $B_0 \not\leq_T A$ and $B_1 \not\leq_T A$.

   Some of the strategies employed in this book are summarized in Fig. 14.14. In the first line of the table, we are building two c.e. sets B_0 and B_1. In remaining four lines of the table, we are given a c.e. set A and want to build a c.e. set B with the listed property (and perhaps with certain other properties).

   What makes the proof of the Density Theorem so delicate is that it simultaneously implements three of these strategies—coding, permitting, and the Friedberg-Muchnik method—while preventing them from interfering with each other. Furthermore, the witness list idea in this chapter seems akin to the length-of-agreement method; so, in a sense, our proof of the Density Theorem uses the entire table.

desired property	strategy
$B_0 \not\leq_T B_1$ and $B_1 \not\leq_T B_0$	Friedberg-Muchnik method
$B \leq_T A$	various types of permitting
$A \not\leq_T B$	length-of-agreement method
$A \leq_T B$	coding
$B \not\leq_T A$	combining the Friedberg-Muchnik method with coding

Fig. 14.14   Some strategies employed in this book

3. In the proof of Theorem 7 (Frieberg-Muchnik), for each requirement we maintained a single witness. In the proof of Theorem 8 (Friedberg-Muchnik below C), for each requirement we maintained up to three witnesses, to facilitate a simple form of permitting. In the proof of Theorem 14 (Density), for each node σ, we maintain an arbitrarily long list of witnesses (n_σ), to facilitate a more complicated form of permitting.
4. The "window of opportunity" for inserting an element clarified the workings of Algorithm 14.3. Might this metaphor help us to understand certain more complicated algorithms?
5. In all tree constructions, it's essential that TP be an infinite path. That result is trivial for finitely branching trees, and rather simple for the infinitely branching tree used in Chap. 12. For the tree in this chapter, the result is still true (it follows from Fact 6), but proving it takes some effort.

   Is the difficulty of proving TP to be an infinite path a useful way to classify the complexity of various tree algorithms? Can it be formalized?

14.6   Designing an Algorithm

Books about algorithms are misleading. They present a problem, then an algorithm to solve it, then a proof of correctness (and then an analysis of the running time, unless the book is about computability theory). You get the impression that the research occurred in that order. However, for complicated algorithms such as Algorithm 14.1, it almost never happens that way. Rather, the algorithm and its proof of correctness are developed together, to some extent. That is, you think about the algorithm in broad strokes, and how the proof might be structured, which leads to refinements of the algorithm, which leads to changes in the proof, necessitating more "wrinkles" in the algorithm. Even more changes in the algorithm occur as you double-check the proof (and perhaps implement and run the algorithm, for the more concrete problems). For example, while trying to verify an early draft of Algorithm 14.1, I saw that

$$\sigma \in TP \implies \liminf_s m_\sigma(s) < \infty$$

(which now follows from Fact 5) was essential to the argument. When trying to prove this, I thought that it would be helpful if no odd number could cause an unrealization; at that time, this was not true, but I made a few changes to the algorithm to make it so (it is now Fact 3).

At various points during the design of Algorithm 14.1, I was faced with a choice, either of which seemed workable. For example, should unrealization take precedence over insertion, or vice versa? If neither alternative was clearly preferable, then I would choose one, and put the other in my "back pocket." Whenever I ran into a difficulty in writing out the verification section, I searched that back pocket (which contained as many as half a dozen items), to see whether any of those ideas would get me past the sticking point. Unfortunately, sometimes a back pocket algorithmic idea would resolve the current verification difficulty but create others.

Algorithm 14.1 is a very different animal from Algorithm 5.1, in terms of the design process.

14.7 Afternotes

Even though this is a book about algorithms, we cannot help but marvel at the partial ordering induced on c.e. sets by Turing reductions. It is dense, according to Theorem 14. On the other hand, the partial ordering on *all* (not necessarily c.e.) sets induced by Turing reductions is not dense. For example, there exists a minimal set, that is, a non-computable, non-c.e. set A such that there is no B such that $\emptyset <_T B <_T A$ (see Chapter V of [Le80]). Thus, the c.e. sets are dense with regard to Turing reductions, whereas sets in general are not. This is surprising, because there are $\aleph_1$ sets in general but only $\aleph_0$ c.e. sets; yet in this sense the smaller family is the dense one.

The Density Theorem was originally proved in [Sa64], using intricate combinatorial arguments; it's not an easy read. Another proof, using index sets (which we have not discussed in this book), appeared in [Ya66b]. Yet another proof, using true stages and hatted restraint functions, appeared in [So87]. A version using a priority tree was sketched in [DH].

Our proof is mostly based on the sketch in [DH], but has some new ideas and is far more detailed. In the tree used by [DH], each node σ has branches

$$(0,\ u),\ (0,\ f),\ (1,\ u),\ (1,\ f),\ \ldots$$

(see Fig. 14.15). The interpretation of the guesses corresponding to the branches is more complicated for their tree than for ours. An algorithm based on their tree would appear to necessitate a more slippery verification argument (if written out in the level of detail seen in this chapter). In priority tree arguments, small changes in the tree can lead to big changes in conceptual simplicity.

Fig. 14.15 Downey-
Hirshfeldt's tree for the proof
of the Density Theorem

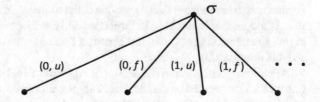

14.8   Exercises

1. Prove that (14.2) implies (14.1).
2. Prove that $A \leq_T B_0$ and $A \leq_T B_1$.
3. Fix e. As we stated in Sect. 14.2,

   ($\exists \sigma$ on level $2e$ of the tree)[σ has a permanent star witness] $\implies$ R_{2e} is met.

   Must the converse be true? In other words, is it possible that R_{2e} is met but

   ($\forall \sigma$ on level $2e$ of the tree)[σ has no permanent star witness]?

4. If $m_\sigma(s)$ converges, can σ perform infinitely many insertions?
5. (a) Prove that B_0 and B_1 each contains infinitely many odd numbers.
   (b) Must they each contain infinitely many even numbers?
6. Suppose that $n_\sigma(i)$ is undefined at the start of a $(\sigma \,\widehat{\ }\, i)$-stage. What are the possible values of i?
7. Fix i. Can there be more than one node σ such that $n_\sigma(i)$ gets a fresh number by line 3 of the unrealization subroutine?
8. Prove that the function $entryIntoC(i)$ is not computable (recall that the definition of a function being computable is in Chap. 3).
9. Suppose that we modify Algorithm 14.1 so that when a number i enters C, witness $n_\sigma(i)$ is immediately put into B_1 (if $|\sigma|$ is even) or into B_0 (if $|\sigma|$ is odd), for each σ such that $i < m_\sigma(i)$. In other words, we immediately act once witness $n_\sigma(i)$ is C-permitted, without waiting for the next σ-stage. Would the algorithm still work (that is, would B_0 and B_1 still satisfy all of the requirements)?
10. Find the flaw in the following argument:
    "Claim:" Suppose that we remove Case 2 from the **switch** statement in Algorithm 14.1 (thus, $B_0 \cup B_1$ would contain no odd numbers). Then the R-requirements would still be met.
    "Proof:" Fix e. Let σ be the level $2e$ node of TP, and let

    $$k = \liminf_s m_\sigma(s).$$

    Then k is a valid witness for R_{2e}. Analogously, R_{2e+1} is met. $\square$

This argument must have a flaw, because without odd numbers, B_0 and B_1 would be equal, contradicting the incomparability of B_0 and B_1.

11. Let σ, e, and k be such that $|\sigma| = 2e$ and $\sigma ^\frown k \in TP$. For each i, let $f(i)$ denote the final value of $n_\sigma(i)$, if such a value exists; otherwise, we write $f(i) \uparrow$.

   (a) Might there exist $i < k$ such that $\Phi_e^{B_0}(f(i)) \neq 0$?
   (b) Might $\Phi_e^{B_0}(f(k)) = 0$?
   (c) Might there exist $i > k$ such that $\Phi_e^{B_0}(f(i)) = 0$?

12. Let $\sigma \in TP$. Might σ perform infinitely many insertions?
13. (a) Does it matter whether insertion takes precedence over realization?
   (b) Does it matter whether unrealization takes precedence over insertion?
   (c) Does the precedence of initialization, relative to that of unrealization, insertion, and realization, matter?
14. Let $\sigma \in TP$. Prove that

$$\lim_s m_\sigma(s) \text{ exists} \iff \sigma \text{ takes the default action during all but finitely many visits to } \sigma.$$

15. In the proof of Lemma 1, must n be a valid witness for R_{2e}?
16. Describe a way to compute $cannot\,Visit\,Node\xi(s_2)$ (which is used in Algorithm 14.3), given s_2, in a finite amount of time, using an oracle for A.
17. Near the end of the proof of Proposition 5, is written:

   By Fact 3, b is even. Hence, there exists $s \geq s_2$ such that

$$b = 2a_s \leq \varphi(\sigma, i)[s_2].$$

   Explain why $s \geq s_2$.
18. Is it possible that

$$(\exists m)(\forall \sigma)(\forall s)\big[m_\sigma(s) \leq m\big]?$$

Chapter 15
The Theme of This Book: Delaying Tactics

There is no sense of urgency in the construction of c.e. sets.[1] We do not care how much time a stage takes, as long as it is finite. Nor do we care how many stages we must wait for an event to happen, as long as that, too, is a finite number.

When all finite amounts of time are regarded as the same, delaying tactics do not hurt, but they might help. Indeed, many of the key algorithmic ideas in this book are delaying tactics; as examples:

1. When permitting is used, a witness must wait for permission to enter the set under construction.
2. A node σ in a priority tree is idle except during σ-stages.
3. During each stage s such that $TP_s <_L \sigma$, a node σ in a priority tree gets re-initialized or canceled in some way, thereby (typically) wiping out any progress that it has already made toward meeting its associated requirements. Fortunately, if $\sigma \in TP$ then there are infinitely many σ-stages, but only finitely many such cancellations.
4. In Chaps. 11 and 12, we use the node-specific computation $\Phi_\sigma^B(k)[s]$, rather than $\Phi_e^B(k)[s]$, to define the length of agreement. In other words, we pay attention to the computation $\Phi_e^B(k)[s]$ only if it is σ-believable. This has the side effect of slowing down (but not stopping) the growth of the length of agreement, if $\sigma \in TP$.
5. In Chap. 14, the star witness concept has the (harmless) side effect of slowing down but not stopping the growth of a witness list.
   Likewise in Chap. 14, both unrealization and insertion take precedence over realization. This, too, might slow down (but not stop) the growth of a witness list.

Both of these delaying tactics are discussed in Note 7 of Sect. 3.3 of Chap. 14.

---

[1] Hence there is no mention of fast data structures in this book.

© The Author(s), under exclusive license to Springer Nature Switzerland AG 2023   163
K. J. Supowit, *Algorithms for Constructing Computably Enumerable Sets*,
Computer Science Foundations and Applied Logic,
https://doi.org/10.1007/978-3-031-26904-2_15

Appendix A
A Pairing Function

Let

$$\langle i,\ j \rangle =_{def} \frac{1}{2}(i + j)(i + j + 1) + j.$$

We are interested in $\langle i,\ j \rangle$ because it is a computable, one-to-one correspondence from $\omega \times \omega$ to ω (it is not unique in that regard).

A minor technical detail arises here: we have not yet defined what it means for a function of two (or more) variables to be computable. One way to do this is to modify our definition of the input to a Turing machine. In particular, suppose that we want to give the machine a finite sequence of inputs $k_1, k_2, \ldots, k_r$. Then, when the machine starts, the binary encoding of k_1 appears on the tape starting at cell 1, followed by one blank symbol (B), followed by the binary encoding of k_2, followed by one B, and so forth, up through the binary encoding of the last input (k_r); the rest of the tape holds nothing but B's.

Note that $\langle i,\ j \rangle$ is monotonically increasing in both i and j. We can view $\langle i,\ j \rangle$ as a matrix, as in Fig. A.1.

Fig. A.1 The pairing function $\langle i,\ j \rangle$

$i \backslash j$	0	1	2	3	4	5	$\cdots$
0	0	2	5	9	14	20	$\cdots$
1	1	4	8	13	19	$\cdots$	
2	3	7	12	18	$\cdots$		
3	6	11	17	$\cdots$			
4	10	16	$\cdots$				
5	15	$\cdots$					

© The Editor(s) (if applicable) and The Author(s), under exclusive license to Springer Nature Switzerland AG 2023
K. J. Supowit, *Algorithms for Constructing Computably Enumerable Sets*,
Computer Science Foundations and Applied Logic,
https://doi.org/10.1007/978-3-031-26904-2

Bibliography

1 Books

[AK]   Ash CJ, Knight J (2000) Computable structures and the hyperarithmetical hierarchy, Studies in Logic and the Foundations of Mathematics, vol 144 Elsevier, Amsterdam

[CLRS]   Cormen TH, Leiserson CE, Rivest RL, Stein C (2009) Algorithms, 3rd ed. MIT Press

[Co]   Cooper SB (2004) Computability theory. Chapman and Hall/CRC Mathematics, London, New York

[Coh]   Cohen PJ (2008) Set theory and the continuum hypothesis. Dover, Minneola, NY

[DH]   Downey RG, Hirshfeldt DR (2010) Algorithmic randomness. Springer, New York

[Ku]   Kunen K (1983) Set theory: an introduction to independence proofs. North Holland

[Le80]   Lerman M (1980) Degrees of unsolvability. Perspectives in Mathematical Logic. Springer

[Le10]   Lerman M (2010) A framework for priority arguments. Cambridge University Press

[Ma]   Martin J (2002) Introduction to languages and the theory of computation, 3rd ed. McGraw-Hill

[Od]   Odifreddi P (1989) Classical recursion theory. North-Holland, Amsterdam

[Ro]   Rogers H Jr (1967) Theory of recursive functions and effective computability. McGraw-Hill, New York

[Si]   Sipser M (2012) Introduction to the theory of computation, 3rd ed. Cengage Learning

© The Editor(s) (if applicable) and The Author(s), under exclusive license to Springer Nature Switzerland AG 2023
K. J. Supowit, *Algorithms for Constructing Computably Enumerable Sets*,
Computer Science Foundations and Applied Logic,
https://doi.org/10.1007/978-3-031-26904-2

[So87]   Soare RI (1987) Recursively enumerable sets and degrees. Perspectives
           in Mathematical Logic. Springer, Heidelberg
[So16]   Soare RI (2016) Turing computability. Springer
[Va]   Vaught RL (2001) Set Theory, 2nd ed. Birkhauser, Boston

2 Articles

[CGS]   Cholak P, Groszek M, Slaman T (2001) An almost deep degree. J Symbol
           Logic 66(2)
[Cs]   Csima BF (2021) Understanding frameworks for priority arguments in
           computability theory. Assoc Symbol Logic Invited Add Joint Math Meet
[De]   Dekker JCE (1954) A theorem on hypersimple sets. Proc Am Math Soc
           5:791–796
[DS]   Downey R, Stob M (1993) Splitting theorems in recursion theory. Ann
           Pure Appl Logic 65(1):1–106
[Fr57]   Friedberg RM (1957) Two recursively enumerable sets of incomparable
           degrees of unsolvability. Proc Nat Acad Sci 43:236–238
[Fr58]   Friedberg RM (1958) Three theorems on recursive enumeration. J Symbol
           Logic 23:308–316
[Ku]   Kucera A (1986) An alternative, priority-free solution to Post's problem.
           In: Lecture Notes in Computer Science, vol 233. Springer, pp 493–500
[La66]   Lachlan AH (1966) Lower bounds for pairs of recursively enumerable
           degrees. In: Proceedings of the London Mathematical Society, vol 16, pp
           537–569
[La79]   Lachlan AH (1979) Bounding Minimal Pairs. J Symbol Logic 44, No 4
[Mi]   Miller A (2007) Lecture Notes in Computability Theory (available as a
           pdf online). University of Wisconsin
[Mo]   Montalbán A (2014) Priority arguments via true stages. J Symbol Logic
           79, No 4:1315–1355
[Mu]   Muchnik AA (1956) On the unsolvability of the problem of reducibility
           in the theory of algorithms. Doklady Akademii Nauk SSSR, NS, vol 108,
           pp 194–197 (Russian)
[Po]   Post EL (1944) Recursively enumerable sets of positive integers and their
           decision problems. Bull Am Math Soc 50:284–316
[Sa63]   Sacks GE (1963) On the degrees less than $\mathbf{0}'$. Ann Math 77, No 2:211–231
[Sa64]   Sacks GE (1964) The recursively enumerable degrees are dense. Ann
           Math 80, No 2:300–312
[Tr]   Trahtenbrot BA (1970) On autoreducibility. Doklady Akademii Nauk
           SSSR 192:1224–1227. An English translation is in Soviet Math 11, No
           3:814–817

[Ya66a]   Yates CEM (1966) A minimal pair of recursively enumerable degrees.
              J Symbol Logic 31:159–168
[Ya66b]   Yates CEM (1966) On the degrees of index sets. Trans Am Math Soc
              121:309–328

Solutions to Selected Exercises

Chapter 3

1. (b) No, H (or any other c.e.n. set) is a counter-example.

   (c) No. There are only $\aleph_0$ c.e. sets, and only $\aleph_0$ complements of c.e. sets (both because there are only $\aleph_0$ Turing machines). However, there are $\aleph_1$ subsets of ω.

2. (a) Yes, because A is infinite (because it is non-computable) and therefore has $\aleph_1$ subsets, but there are only $\aleph_0$ c.e. sets.

   (b) Yes. Let $a_0, a_1, \ldots$ be a standard enumeration of A. Let B be constructed by the following algorithm:

   > $B \leftarrow \emptyset$.
   > **for** $i \leftarrow 1$ **to** ∞
   >   **if** $a_i > \max\{a_0, a_1, \ldots, a_{i-1}\}$
   >   put a_i into B.

   Then B is infinite and computable.

3. (a) No. Let P be non-computable, and let

   $$A = \{2p : p \in P\} \cup \{2p+1 : p \in P\}.$$

   Then

   $$B_0 = \{2p : p \in P\}$$

   and

   $$B_1 = \{2p+1 : p \in P\}.$$

© The Editor(s) (if applicable) and The Author(s), under exclusive license to Springer Nature Switzerland AG 2023
K. J. Supowit, *Algorithms for Constructing Computably Enumerable Sets*,
Computer Science Foundations and Applied Logic,
https://doi.org/10.1007/978-3-031-26904-2

Set B_0 is not computable, because otherwise we could determine whether a given p is in P as follows:

if $2p \in B_0$
output("yes")
else output("no").

Likewise B_1 is not computable, because otherwise we could determine whether a given p is in P as follows:

if $2p + 1 \in B_1$
output("yes")
else output("no").

(b) Yes, because otherwise A would be the union of two computable sets, and hence computable.

Chapter 4

1. The **while** loop would indeed halt for each s. To prove this, fix s. There is a Turing machine that accepts a_s but nothing else; let j be its index. Thus, $W_j = \{a_s\}$. Let $z > j$ be such that $\Phi_j(a_s)[z] \downarrow$. During stage s, if the **while** loop eventually sets t to z, then *assigned* is set to TRUE.

   I don't see how every requirement would be met. Fix some e such that W_e is infinite. Consider the requirement $R_{e,1}$. Perhaps for each s such that $a_s \in W_e$, some requirement of weaker priority than $R_{e,1}$ puts a_s into B_0. In particular, let z be the least number such that $a_s \in W_e[z]$. There might be some $e' > e$ and some $z' < z$ such that
   $$a_s \in W_{e'}[z'] \text{ and } W_{e'}[z'] \cap B_0 = \emptyset$$

   when $t = z'$ during stage s; if so, then a_s is put into B_0.

3. Our requirements are now
$$R_{e,i} : W_e \neq \overline{B_i}$$

   for each $e \in \omega$ and each $i \in \omega$ (whereas before it was just $i \in \{0, 1\}$).
   The priorities among the requirements are defined as follows: $R_{e,i} \prec R_{e',i'}$ if and only if
   $$\langle e, i \rangle < \langle e', i' \rangle.$$

   The algorithm is now:

    for $s \leftarrow 0$ **to** ∞
        $B_s \leftarrow \emptyset$.
        assigned $\leftarrow$ FALSE.
        for $k \leftarrow 0$ **to** s
            $e, i \leftarrow$ the unique numbers such that $k = <e, i>$.
            if (not *assigned*) and $a_s \in W_e[s]$ and $W_e[s] \cap B_i = \emptyset$
                // Requirement $R_{e,i}$ *acts*.
                Put a_s into B_i.
                assigned $\leftarrow$ TRUE.
    if not *assigned*
        Put a_s into B_0.   // Or put a_s into B_1, it doesn't matter.

The rest of the proof is very much like the proof of Theorem 4.

Incidentally, here's an incorrect answer that some students have given for this problem: "Use Theorem 4 repeatedly, to obtain, by induction, a partition of A into infinitely many c.e.n. sets." It's incorrect because what it actually proves is the weaker statement:

$$(\forall k)[A \text{ can be partitioned into } k \text{ c.e.n. sets.}]$$

5. No, because B_0 and B_1 are c.e.

Chapter 5

5. Let $\sigma_0, \sigma_1, \ldots$ denote the finite-length binary strings in lexicographic order. Consider the following algorithm, which takes n as input:

    ALGORITHM Y
        for $s \leftarrow 0$ **to** ∞
            for $i \leftarrow 0$ **to** s
                if $\beta_{2e} \prec \sigma_i$
                    if the computation $\Phi_e^{\sigma_i}(n)$ halts within s steps, and $\Phi_e^{\sigma_i}(n)=1$
                        output(σ_i).
                        halt.

The number e and the string β_{2e} are "hard-coded" into Algorithm Y.

Now, consider another algorithm, which is given n as input, and has H as an oracle:

    ALGORITHM Z
        $e' \leftarrow$ the index of a Turing machine that executes Algorithm Y.
        if $\langle e', n \rangle \in H$ // The oracle for H is used here.
            $\sigma \leftarrow$ the output of Algorithm Y on input n.
            output(σ)
        else output("no").

Algorithm Z solves the problem for (5.1). The argument for (5.2) is analogous, substituting α_{2e+1} for β_{2e}.

7. It is not clear that a Turing machine without an oracle can determine whether (5.1) or (5.2) are true; so, even though the elements of A and of B are enumerated in ascending order, they are not necessarily *computably* enumerated.

Note: one might think that this constitutes a proof that A and B are non-c.e., but such an argument would be invalid, because it would not rule out the possibility that A and B could be computably enumerated by an algorithm very different from Algorithm 5.1.

8. If the α strings were constructed by Algorithm 5.1, then we would have

$$|\alpha_0| = 0,$$

$$|\alpha_1| = 1,$$

and

$$|\alpha_3| = |\alpha_2| + 1.$$

Indeed, for each e, we would have

$$|\alpha_{2e+1}| = |\alpha_{2e}| + 1$$

and

$$|\beta_{2e+2}| = |\beta_{2e+1}| + 1.$$

10. (a) Use Exercise 9(b), and induction.
    (b) Let $B = \{ \langle i, k \rangle : k \in A_i \}$. Then $A_i \leq_T B$ for each i. Informally, B encodes all of the information about the A_i.

    To show $(\forall i)[B \not\leq_T A_i]$, assume for a contradiction that there were an i such that $B \leq_T A_i$. Then

    $$A_{i+1} \leq_T B \leq_T A_i,$$

    contradicting $A_i <_T A_{i+1}$.

Chapter 6

1. Yes. Let $u = \varphi_e^D(k)[s]$. By the definition of permanence, $\Phi_e^D(k)[s] \downarrow$ and

$$D_s \restriction u = D \restriction u.$$

Up through step s, during the computation $\Phi_e^D(k)$, the oracle head never moves to the right of cell u. Hence, computations $\Phi_e^D(k)$ and $\Phi_e^D(k)[s]$ are identical in every way.

2. No, because $D - D_x$ might contain a number less than or equal to $\varphi_e^D(k)[x]$.

3. No. Suppose that

$$(\forall z \geq y)\big[\, d_{z+1} \leq \varphi_e^D(k)[z] \,\big].$$

It might be that

$$(\forall z \geq y)\big[\, \varphi_e^D(k)[z] < \varphi_e^D(k)[z+1] \,\big]$$

even though

$$(\forall z \geq y)\big[\, \varphi_e^D(k)[z] = 5 \,\big].$$

Thus, for each z, the computation $\Phi_e^D(k)[z]$ is not permanent. Therefore $\Phi_e^D(k)\uparrow$, by the contrapositive of the Permanence Lemma.

Chapter 7

1. Consider two even numbers $i < j$. If R_i and R_j share a witness, which is put into A because R_j acts, then that witness might injure R_i. Thus, we would no longer be able to argue that a requirement can be injured only by the action of a stronger priority requirement.
   The analogous problem could arise if two odd-numbered requirements shared a witness.
   I don't see a difficulty with an even-numbered requirement sharing a witness with an odd-numbered requirement.
2. (a) One.
   (b) Two.
   (c) An upper bound of 2^j can be derived by assuming that R_i acts the maximum number of times, for each $i < j$.
   (d) If j is even then the action of R_j puts an element into A, which cannot injure an even-numbered requirement (because those try to keep elements out of B). Likewise, if j is odd then the action of R_j cannot injure an odd-numbered requirement. Hence, line 9 of Algorithm 7.1 can be replaced by:

   > **if** j is even
   >   *Initialize*(k) for each odd k such that $j < k \leq s$
   > **else** *Initialize*(k) for each even k such that $j < k \leq s$

   The Fibonacci numbers are usually defined as $F_0 = 0$, $F_1 = 1$, and $F_j = F_{j-1} + F_{j-2}$ for all $j \geq 2$.
   There are two well-known identities that regard sums of all even- or odd-indexed Fibonacci numbers (both of which can be proved by a standard inductive argument):

   $$\sum_{i=1}^{j} F_{2i} = F_{2j+1} - 1 \quad \text{and} \quad \sum_{i=0}^{j-1} F_{2i+1} = F_{2j}.$$

   Define $a(j)$ to be the maximum number of times R_j can act, then we have $a(0) = 1$, $a(1) = 2$, and in general

$$a(j) = \begin{cases} 1 + \sum_{i=0}^{(j-1)/2} a(2i), & \text{if } j \text{ is odd} \\[2mm] 1 + \sum_{i=0}^{j/2-1} a(2i+1), & \text{if } j \text{ is even.} \end{cases}$$

We will prove by induction that

$$a(j) = F_{j+2}.$$

Our base cases have been established, so assume that $a(i) = F_{i+2}$ for all $i < k$. Then we have two calculations depending on whether k is odd or even. If it is odd then

$$a(k) = 1 + \sum_{i=0}^{(k-1)/2} a(2i)$$

$$= 1 + \sum_{i=0}^{(k-1)/2} F_{2i+2}$$

$$= 1 + \sum_{i=1}^{(k+1)/2} F_{2i}$$

$$= 1 + (F_{2\cdot(k+1)/2+1} - 1)$$

$$= F_{k+2}.$$

If it is even then

$$a(k) = 1 + \sum_{i=0}^{k/2-1} a(2i+1)$$

$$= 1 + \sum_{i=0}^{k/2-1} F_{2i+3}$$

$$= 1 + \sum_{i=1}^{k/2} F_{2i+1}$$

$$= 1 + (F_{2\cdot(k/2+1)} - 1)$$

$$= F_{k+2}.$$

Thus $a(k) = F_{k+2}$, so by induction $a(j) = F_{j+2}$ for all j. It is known that

$$F_j = \frac{\left(\frac{1+\sqrt{5}}{2}\right)^j - \left(\frac{1-\sqrt{5}}{2}\right)^j}{\sqrt{5}}$$

and so

$$F_j \in \Theta\left(\left(\frac{1+\sqrt{5}}{2}\right)^j\right).$$

Because

$$\frac{1+\sqrt{5}}{2} \approx 1.618\cdots$$

this is asymptotically better than 2^j.

This solution was provided by Oscar Coppola.

Can the algorithm be further modified so as to reduce this upper bound even more? I don't know.

6. For each e, define the requirement:

$$R_e : (\exists n)\left[\chi_A(n) \neq \Phi_e^{A-\{n\}}(n)\right].$$

Let n_e denote the witness for R_e (n_e is a variable, as in the Friedberg-Muchnik proof). Say that R_e *needs attention* at stage s if

$$n_e \notin A_s \quad \text{and} \quad \Phi_e^{A-\{n_e\}}(n_e) = 0.$$

We define A as the union of finite sets

$$A_0 \subseteq A_1 \subseteq \cdots.$$

Here's the algorithm, where procedure *Initialize* works exactly as it did in our proof of the Friedberg-Muchnik Theorem:

$A \leftarrow \emptyset.$
$n_0 \leftarrow 0.$ // This is equivalent here to *Initialize*(0).
for $s \leftarrow 1$ **to** ∞
    Initialize(s).
    for $j \leftarrow 0$ **to** s
        if R_j needs attention
            // R_j acts.
            Put n_j into A.
            Initialize(k) for each k such that $j < k \leq s$.

The set A is c.e. because each stage can be performed in a finite amount of time.

To show that each R_e is met, we could use an inductive proof that is analogous to that of Lemma 7.1.

Chapter 8

1. Suppose that R_j acts during a stage $z > w$. This action creates a Type 3 witness for R_j; in other words, $n_j(3)[z + 1] \geq 0$. It also removes the Type 2 witness; in other words, $n_j(2)[z + 1] = -1$. The only way for $n_j(2)$ to become non-negative again is by line 20 of Algorithm 8.1, which cannot happen as long as $n_j(3) \geq 0$ (because of the second condition in line 19). Furthermore, the only way for $n_j(3)$ to reset to -1 after stage z is for R_i to act for some $i < j$, which is impossible because $z > w$.

   Therefore, R_j never acts after stage z, and so R_j acts at most once after stage w.

5. (a) Yes.

   (b) Yes.

Chapter 9

3. Assume (9.7). Then $\Phi_e^{B_i}(p) \downarrow$ and $A(p) \neq \Phi_e^{B_i}(p)$. By the Permanence Lemma, there is a stage s_0 such that the computation $\Phi_e^{B_i}(p)[s_0]$ is permanent. Let $s_1 > s_0$ be such that

$$A_{s_1}(p) = A(p).$$

   Then

$$(\forall s \geq s_1)\big[A_s(p) \neq \Phi_e^{B_i}(p)[s]\big]$$

   and so

$$\max_{s \geq s_1} \ell_{e,i}(s) \leq p.$$

   Therefore

$$\max_s \ell_{e,i}(s) < \infty,$$

   contradicting (9.5).

7. (a) Let $A = C$ be c.e.n. Assuming Theorem 9', A can be partitioned into c.e. sets B_0 and B_1 such that

$$A \not\leq B_0 \quad \text{and} \quad A \not\leq B_1.$$

   By Lemma 9.1, B_0 and B_1 are incomparable.

   (b) *Hint*: Modify the proof of Theorem 9 by defining $\ell_{e,i}$ as the length of agreement between C and $\Phi_e^{B_i}$.

8. *Hint*: Enhance Algorithm 9.1 with C-permitting (as was used in Algorithm 8.1).

Chapter 11

1. Let $B = \emptyset$.

3. Let C be c.e.n.

   (a) Let $A = \{\langle e, 0\rangle : e \in C\}$, and suppose that A is given as input to Algorithm 11.1. Each row of A is finite, and so, by Lemma 11.1,

$$(\forall \sigma \in TP)[\sigma \frown f \in TP].$$

Therefore, TP is computable.

(b) Let $A = \{\langle e, j \rangle : e \in C \text{ and } j \in \omega\}$, and suppose that A is given as input to Algorithm 11.1. Then

$$(\forall e)\big[e \in C \iff A^{[e]} \text{ is full} \iff \sigma \frown \infty \in TP\big],$$

where σ is the level e node of TP. Therefore, given e, we could determine whether $e \in C$ by using an oracle for TP. In other words,

$$C \leq_T TP$$

and hence TP is non-computable.

11. (b) No. This change would make the definitions circular: $restraint_\sigma$ depends on ℓ_σ, which depends on Φ_σ^B, which depends on σ-believability, which would depend on $restraint_\sigma$.

13. Yes, because $s > x$, $p < \ell_\sigma(s)$, and condition (iv) on the choice of x in Part I of Lemma 11.2 together imply

$$\Phi_\sigma^B(p)[s] = \Phi_e^B(p)[s].$$

Chapter 12

2. Yes. Fix e, r, and s. Both A_s and $W_r[s]$ are finite sets; in particular,

$$|A_s| = |W_r(s)| = s + 1.$$

Hence, for sufficiently large k,

$$A(\langle e, k \rangle)[s] = W_r(\langle e, k \rangle)[s] = 0.$$

3. Fix e and r. Suppose that

$$\overline{\ell}_{e,r}(s - 1) > \overline{\ell}_{e,r}(s)$$

for some $s \geq 1$. Then

$$(\exists k < \overline{\ell}_{e,r}(s - 1))\big[A(\langle e, k \rangle)[s] = 1 = W_r(\langle e, k \rangle)[s]\big].$$

Thus, whenever $\overline{\ell}_{e,r}$ decreases, it never regains its former size. Therefore, if $\overline{\ell}_{e,r}$ decreases at some point then it is unimodal; otherwise, it is monotonically increasing.

It is possible that $\overline{\ell}_{e,r}$ is both unimodal and monotonically increasing.

7. (a) The following algorithm differs from Algorithm 11.1 only in that line 8 has
    been changed from "**else** $\tau \leftarrow \tau \frown f$."

    1   $B \leftarrow \emptyset$.
    2   **for** $s \leftarrow 0$ **to** ∞
            // Compute TP_s .
    3        $TP_s \leftarrow \{\lambda\}$.
    4        $\tau \leftarrow \lambda$.
    5   **for** $e \leftarrow 0$ **to** $s - 1$
    6   **if** $\left| A_s^{[e]} \right| > \left| A_{pred_\tau(s)}^{[e]} \right|$
    7                $\tau \leftarrow \tau \frown \infty$
    8   **else** $\tau \leftarrow \tau \frown \left| A_s^{[e]} \right|$.

    9        $TP_s \leftarrow TP_s \cup \{\tau\}$.

            // $B \leftarrow B \cup \{\text{certain elements of } A\}$.
    10   **for** $e \leftarrow 0$ **to** $s - 1$
    11           $\xi \leftarrow$ the level e node of TP_s.
    12   **for** each $a \in A_s^{[e]} - B$ such that $a > Restraint_\xi(s)$
    13   Put a into B.

  (b) No. Because $TP <_L \xi$, there exists $\sigma \in TP$ such that $\sigma \lhd \xi$. The picture
    looks either like Fig. S.1a or b below, for some τ and r, where $e = |\tau|$.
    If $A^{[e]}$ is full (as in Fig. S.1a) then there is an x such that $\left| A_x^{[e]} \right| > r$.
    Node $\tau \frown r$ is never visited after stage x. Hence there are only finitely many
    ξ-stages.
    So assume that $A^{[e]}$ is finite; let $k = \left| A^{[e]} \right|$ (as in Fig. S.1b). Then, because
    $k < r$, there are no ξ-stages.

  (c) Not necessarily. Suppose that $A^{[0]}$ is full, and that $A - A^{[0]}$ is infinite. Then
    there are infinitely many s such that $\infty \in TP_s$, and infinitely many s such
    that $\infty \notin TP_s$. Thus, the claim would fail for $e = 1$.

Chapter 13

3. Yes. If $A_s \cap W_e[s] \neq \emptyset$ then requirement $P_{|\sigma|,A}$ never needs attention during or
   after stage s.
4. The argument in Case 2 of the proof of Proposition 1 within the proof of Lemma
   13.3 would fail.
5. Fix τ. Let
$$S = \{s : TP_s <_L \tau\}.$$

   The value of $n_{\tau,A}$ can change (after its first assignment) only during a stage
   in S. If either $\tau <_L TP$ or $\tau \in TP$ then S would be finite, and so $n_{\tau,A}$ would have
   a final value.
   To prove the converse, suppose $TP <_L \tau$. Let σ be the level $|\tau|$ node of TP. Node
   τ is initialized (with a larger value than it had before) during each σ-stage, of
   which there are infinitely many.

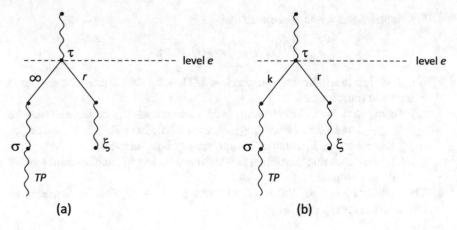

Fig. S.1 Illustration for a solution to Exercise 6(b) of Chap. 12

6. Yes, because t_0 is a σ-expansionary stage, which implies

$$L_\sigma(t_0) = \ell_\sigma(t_0).$$

9. As pointed out near the end of the proof of Lemma 13.3, there is a stage y such that the computation $\Phi_e^A(p)[y]$ is permanent. For each i such that $t_i > y$,

$$\varphi_e^A(p)[t_{i+1}] = \varphi_e^A(p)[t_i] = \varphi_e^A(p)[y].$$

Chapter 14

1. Assume (14.2). We will show $A <_T B_0 <_T C$ (the proof that $A <_T B_1 <_T C$ is analogous). Because $A \leq_T B_0 \leq_T C$, we need prove only that $B_0 \not\leq_T A$ and $C \not\leq_T B_0$. If $B_0 \leq_T A$ then

$$B_0 \leq_T A \leq_T B_1,$$

contradicting $B_0 \not\leq_T B_1$. Likewise, if $C \leq_T B_0$ then

$$B_1 \leq_T C \leq_T B_0,$$

contradicting $B_1 \not\leq_T B_0$.
3. The converse might fail. For example, it might be that the eth OTM never halts, regardless of its input and what is written on its oracle tape. In that case,

$$(\forall k)\left[\Phi_e^{B_0}(k) \uparrow\right]$$

and so R_{2e} would be met, but no node on level $2e$ would ever perform a realization; therefore no node on level $2e$ would ever have a star witness.

4. No. Suppose that k and s_0 are such that

$$(\forall s \geq s_0)[\, m_\sigma(s) = k \,].$$

If $n_\sigma(i)$ is inserted after stage s_0, then $i < k$. Thus, by Fact 1, after stage s_0, node σ inserts at most k times.

5. (a) There must be infinitely many odd numbers in B_0, because otherwise $B_0 \leq_T A$ and hence (since $A \leq_T B_1$), we would have $B_0 \leq_T B_1$. By analogous reasoning, B_1 contains infinitely many odd numbers.

   (b) No, because A might be finite, in which case B_0 and B_1 each contains exactly $|A|$ even numbers.

6. The number i must be 0. If $n_\sigma(i)[s]\uparrow$ and $(\sigma{\,}^\frown i) \in TP_s$ then σ performs an initialization during stage s.

8. For each i,
$$i \in C \iff entryIntoC(i) \geq 0.$$

Hence, if $entryIntoC(i)$ were computable, then, given i, we could ascertain whether $i \in C$ by computing $entryIntoC(i)$, and so C would be computable, contradicting the assumption that $A <_T C$.

10. Here's the flaw: without insertion, there would be no star witnesses, and so the proof of Fact 5, which says

$$\liminf_s m_\sigma(s) < \infty,$$

would fail.

13. (c) No, because whenever the condition for initialization holds, the conditions for unrealization, insertion, and realization do not hold.

15. No. In Case 1.1 of the proof of Lemma 1.1, node σ has a permanent star witness, which could prevent $n_\sigma(k)$ from being realized. Therefore, it is possible that

$$B_1(n) = 0 = \Phi_e^{B_0}(n).$$

16. This is similar to the proof of Proposition 1 (within the proof of Fact 5).
    To determine the truth of $cannotVisitNode\xi(s_2)$, for each σ, i, and i' such that

$$i < i' \quad \text{and} \quad \sigma{\,}^\frown i \preceq \xi \quad \text{and} \quad \sigma{\,}^\frown i' \in TP_{s_2},$$

we need to determine whether

$$\neg(\exists s \geq s_2)[\, 2a_s \leq \varphi(\sigma, i)[s_2] \,].$$

We do this as follows:

> $a \leftarrow 0$.
> $answer \leftarrow$ TRUE.
> **while** $answer =$ TRUE and $2a \leq \varphi(\sigma, i)[s_2]$
> > **if** $a \in A - A_{s_2-1}$ // Here we use the oracle for A.
> > > $answer \leftarrow$ FALSE.
> > $a \leftarrow a + 1$.
> output($answer$).

The **while** loop iterates no more than

$$1 + \varphi(\sigma, i)[s_2]/2$$

times.

18. No; assume for a contradiction that there is an m such that

$$(\forall \sigma)(\forall s)\big[m_\sigma(s) \leq m\big].$$

Let z be such that

$$C_z \upharpoonright m = C \upharpoonright m.$$

Suppose that n is chosen as a fresh witness after stage z. Then n never receives C-permission, and so $n \notin B_0 \cup B_1$. Therefore, B_0 and B_1 are both finite. Hence

$$B_0 \equiv_T B_1,$$

contradicting $B_0 \not\leq_T B_1$ (and $B_1 \not\leq_T B_0$).

Printed in the United States
by Baker & Taylor Publisher Services